The SCIENCE *of the*
EARTH

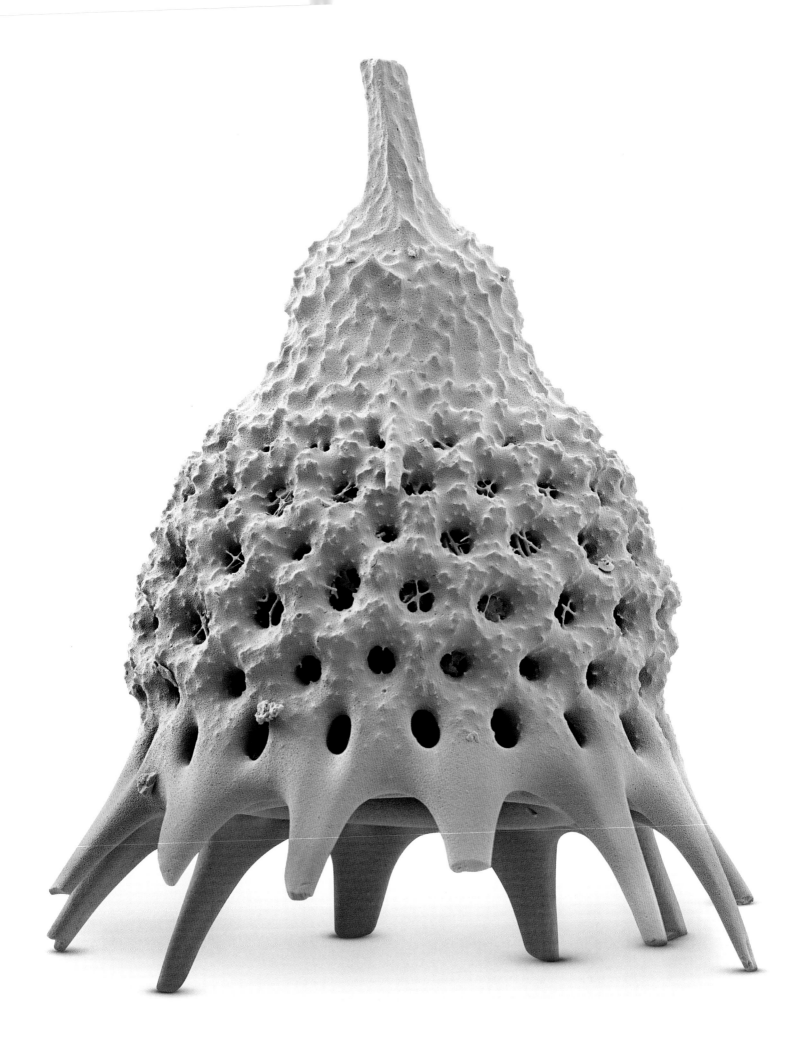

The SCIENCE *of the* EARTH

THE SECRETS OF OUR PLANET REVEALED

Penguin
Random
House

DK LONDON

Senior Editors Peter Frances, Rob Houston
Editors Polly Boyd, Jemima Dunne,
Sarah MacLeod, Cathy Meeus, Steve Setford
Senior Production Controller Meskerem Berhane
Managing Editor Angeles Gavira Guerrero
Associate Publishing Director Liz Wheeler
Publishing Director Jonathan Metcalf

Senior Art Editors Sharon Spencer, Duncan Turner
Illustrator Phil Gamble
Senior Jacket Designer Akiko Kato
Jacket Design Development Manager Sophia MTT
Managing Art Editor Michael Duffy
Art Director Karen Self
Design Director Phil Ormerod

DK DELHI

Senior Editor Dharini Ganesh
Project Editor Priyanjali Narain
Senior Picture Researchers Deepak Negi,
Surya Sankash Sarangi
Senior Jacket Designer Suhita Dharamjit
Senior Jacket Coordinator Priyanka Sharma
Picture Research Manager Taiyaba Khatoon
Senior Managing Editor Rohan Sinha
Editorial Head Glenda R Fernandes

Senior Art Editor Mahua Mandal
Project Art Editor Anjali Sachar
Art Editors Nobina Chakravorty, Debjyoti Mukherjee
DTP Designer Vikram Singh
Production Editor/Senior DTP Designer Vishal Bhatia
Managing Art Editor Sudakshina Basu
Pre-production Manager Balwant Singh
Production Manager Pankaj Sharma
Design Head Malavika Talukder

First published in Great Britain in 2022 by
Dorling Kindersley Limited
DK, One Embassy Gardens, 8 Viaduct Gardens,
London SW11 7BW

The authorised representative in the EEA is
Dorling Kindersley Verlag GmbH. Arnulfstr. 124,
80636 Munich, Germany

Copyright © 2022 Dorling Kindersley Limited
A Penguin Random House Company
10 9 8 7 6 5 4 3 2 1
001–326803–Oct/2022

A CIP catalogue record for this book
is available from the British Library.
ISBN: 978-0-2415-3643-8

Printed in the United Arab Emirates

For the curious
www.dk.com

Contributors

Philip Eales studied physics and remote sensing at University College, London. As well as writing about Earth and space science, he runs a computer graphics company specializing in the visualization of astronomical and geographical data and phenomena.

Gregory Funston is a Canadian palaeontologist who studied at the University of Alberta and is now a postdoctoral fellow at the University of Edinburgh. His research on dinosaurs and mammals has taken him to fossil sites around the world.

Derek Harvey is a naturalist with a particular interest in evolutionary biology, who studied Zoology at the University of Liverpool. He has taught a generation of biologists, and has led student expeditions to Costa Rica, Madagascar, and Australasia.

Anthea Lacchia is a writer and journalist based in Ireland. She mainly writes about science and nature, and has a PhD on the study of fossil ammonoids – extinct relatives of squid and cuttlefish.

Dorrik Stow is a geologist, oceanographer, and author of over 300 papers and books. He is Emeritus Professor at Heriot-Watt University in Edinburgh, Distinguished Professor at the China University of Geoscience in Wuhan, and a Leverhulme Emeritus Fellow.

Half-title page Geode with amethyst and calcite crystals
Title page Coloured scanning electron micrograph of the fossil radiolarian *Anthocyrtidium ligularia*
Above Rhyolite and geothermal deposits at Laugahraun lava field, Iceland
Contents page Flies, borer beetles, and worker ants preserved in amber from the Dominican Republic

Consultants

David Holmes is a geographer who studied at Leeds University, UK, and has a first degree in physical geography and a master's degree in environmental science. He is a Fellow of the Royal Geographical Society in London and is the author of several well-known geography school textbooks.

Cally Oldershaw is former curator of gemstones at the Natural History Museum and Chair of the Gemmological Association of Great Britain. She is an Earth science educational consultant and lecturer, and the author of numerous books and articles.

Douglas Palmer is an Earth science writer who has written and contributed to numerous books, especially on palaeontology, which he originally lectured on at Trinity College Dublin. He also works at Cambridge University's Sedgwick Museum.

Kim Dennis-Bryan is a zoologist who began her career studying fossil fish at London's Natural History Museum, before becoming an Open University lecturer specializing in natural sciences. She has written for and consulted on many science books, including DK's *Animal*, *Ocean*, and *Prehistoric Life*.

MIX
Paper | Supporting
responsible forestry
FSC™ C018179

This book was made with Forest Stewardship Council™ certified paper – one small step in DK's commitment to a sustainable future. For more information go to www.dk.com/our-green-pledge

contents

planet Earth

12 interstellar origins
14 the formation of the Solar System
16 meteorites
18 Tnorala
20 the Moon
22 our place in the Solar System
24 days and seasons
26 Earth's rotation
28 origin of the oceans
30 the continents form
32 the ages of Earth

Earth's materials

36 crystal structure
38 crystal habits
40 reflecting light
42 from hard to soft
44 native elements
46 Earth's metals
48 varieties of quartz
50 feldspars
52 altered minerals
54 mineral associations
56 crystalline cavities
58 gemstones
60 organic minerals
62 the rock cycle
64 igneous rocks of the oceans
66 igneous rocks of the continents
68 El Capitan and Half Dome
70 rocks from lava
72 lava
74 fine-grained rocks
76 sandstone
78 cemented fragments
80 chemical deposition
82 limestone
84 making maps of rocks
86 dynamic and burial metamorphism
88 regional metamorphism
90 contact metamorphism
92 layers of rock
94 the Grand Canyon
96 the soil layer
98 properties of water
100 frozen water
102 avalanches

dynamic Earth

106 Earth's structure
108 satellites and Earth science
110 tectonic plates
112 plate boundaries
114 colliding plates
116 diverging plates
118 Earth's folds
120 mountain-building
122 understanding mountain-building
124 the Himalayas
126 lines of fracture
128 earthquakes and tsunamis
130 measuring earthquakes
132 shaking ground
134 igneous intrusions
136 Shiprock
138 volcanoes
140 volcanic eruptions
142 hotspots
144 geothermal features
146 hot springs
148 erosion by water
150 erosion by wind
152 erosion by ice
154 weathering
156 depositing sediment
158 rivers
160 waterfalls
162 deltas and estuaries
164 lakes
166 karst landscapes
168 caves
170 glaciation
172 Glacier Bay
174 meltwater

ocean and atmosphere

178 ocean chemistry
180 ocean circulation
182 upwelling and plankton blooms
184 deep currents
186 tides
188 oceanic waves
190 frozen sea
192 ice shelves and icebergs
194 where land meets the sea
196 sea-level change
198 shallow seas
200 seafloor mapping
202 open ocean waters
204 patterns on the seafloor
206 ocean tectonics
208 islands in the ocean
210 Earth's atmosphere
212 aurorae
214 wind
216 atmospheric imaging
218 weather systems
220 tropical cyclones
222 types of clouds
224 the water cycle
226 falling rain
228 Mount Wai'ale'ale
230 snow
232 thunderstorms
234 Tornado Alley
236 electricity in the air
238 sprites, jets, and elves
240 ice core analysis
242 natural climate change

living planet

246 the biosphere

248 millions of species

250 early life

252 life transforms Earth

254 life branches out

256 Cambrian explosion

258 making reefs

260 age of fish

262 invading land

264 succession

266 forests form

268 fossilization

270 coal forms

272 oxygen-fuelled giants

274 walking on land

276 the great dying

278 trilobite diversity

280 surviving drought

282 scanning fossils

284 becoming warm-blooded

286 Red Deer River Valley

288 taking to the air

290 giants of the land

292 insect pollinators

294 death of the dinosaurs

296 fossilized community

298 the age of mammals

300 grasslands

302 life on the move

304 surviving ice

306 permafrost

308 boreal forest

310 molecular evidence from fossils

312 megafauna

314 Forest of Borth

316 shaped by humans

318 Danum Valley

320 glossary

326 index

334 acknowledgments

foreword

Initially molten, later frozen, once watery and landless, a planet without a sky or clouds – it's hard to imagine our Earth without sunshine or rain, our beautiful spinning sphere entirely without life. That is not because of the shortcomings in our scientific understanding of its history, it's down to the complexity of factors involved and, critically, our perception of time. Time is tricky to get a measure of.

The fiery ball we now call home formed 4.6 billion years ago. Life appeared at least 3.77, maybe even 4.41, billion years ago. All life descends from the Last Universal Common Ancestor (LUCA), an organism from which every living species alive today has evolved – yes, every plant, animal, and fungal organism is related to the one single-celled life form that would not have looked too different from a small, modern-day bacterium. We have no fossils of LUCA, but we think that it appeared in the hot waters of deep-sea ocean vents 4.28 billion years ago. It is a big idea, complete with very big numbers! So, how do we begin to gather what we know about Earth's complex past and use it to appreciate how the world exists now?

This remarkable and beautiful book offers us that chance. By patiently breaking the big ideas into smaller features about the fundamentals of physics, chemistry, and biology, we can figure out the processes that once formed our world and continue to shape it today. From crystals to tornados and fossils to volcanoes, each page offers exciting insights into what makes Earth work. It is the most exciting story ever told, because without its myriad chapters and all its twists and turns, we simply would not be here today... and it's been quite a journey.

CHRIS PACKHAM
NATURALIST, BROADCASTER, WRITER,
PHOTOGRAPHER, AND CONSERVATIONIST

SANDSTONE RIDGES ON NORWAY'S VARANGER PENINSULA

planet Earth

Earth came into existence about 4.6 billion years ago as a hot, rocky body orbiting the newly formed Sun. Its position in the Solar System makes it unique among the planets in having liquid water on its surface. The continued outward transfer of heat from its interior has also kept it geologically active, driving constant motion and recycling of its outer layers.

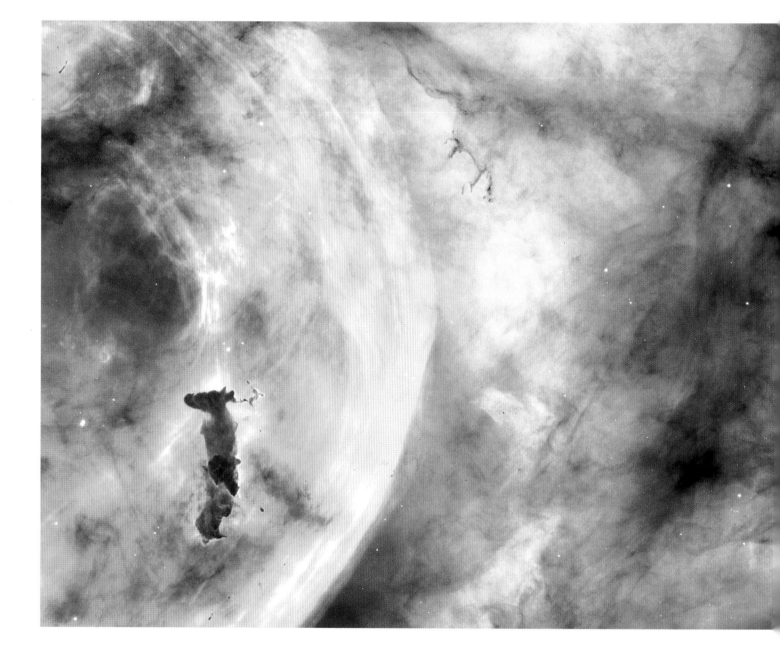

interstellar origins

With three exceptions, every element on Earth was forged long ago in the furnace of a star. The exceptions – hydrogen, helium, and lithium – were created when the Universe was born. The heat and pressure in a star's core are thought to be high enough to fuse lighter elements to create heavier ones, up to iron with an atomic weight of 56. Elements heavier than iron, such as lead, are created by the even more powerful forces unleashed when a dying star explodes as a supernova. The blast scatters all the elements created by the star into interstellar space, seeding the next generation of stars.

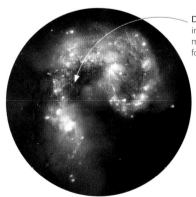

Dust glows red in the light of millions of newly formed stars

Star trigger
When galaxies collide, new waves of star formation are triggered. The Antennae galaxies have been colliding for about 100 million years, throwing clouds of gas and dust together.

THE MILKY WAY

Our home galaxy, the Milky Way, is a barred spiral galaxy containing 100–400 billion stars. Its two major spiral arms extend from the ends of a central bulge densely populated with old stars. As the galaxy rotates, the interstellar gas and dust become compressed, producing areas of star formation. Our Solar System lies in a minor arm called the Orion Spur, a little over halfway between the galaxy's core and its outer arm.

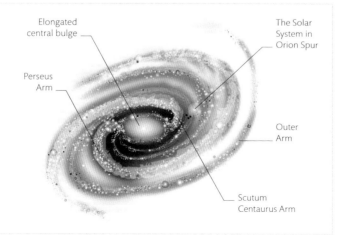

Elongated central bulge

Perseus Arm

The Solar System in Orion Spur

Outer Arm

Scutum Centaurus Arm

STRUCTURE OF THE MILKY WAY

Molecular clouds

The material expelled from stars accumulates as clouds of gas and dust. Consisting mostly of hydrogen molecules, these are known as molecular clouds. The Keyhole Nebula (above, left) is an example of a molecular cloud. Molecular clouds can contract through their own gravity or an external trigger to form dense globules, such as "the Caterpillar" (above, right), that give birth to new stars.

Protoplanetary disc

A disc of dust, called a protoplanetary disc, can be seen swirling around the young star HL Tauri in this image captured by a radio telescope. HL Tauri is 450 light-years from Earth in the constellation Taurus. Bright and dark circles within the disc show where dust is coalescing to form planets.

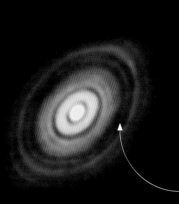

Dark patches indicate where planets may be forming within the disc

The Sun

Accounting for 99 per cent of the mass of the Solar System, the Sun is a yellow dwarf star, about halfway through its life. It has a diameter of about 1.4 million km (865,000 miles). The Sun is the most important source of light and heat for life on Earth. Its energy comes from nuclear fusion in its core, where the temperature reaches 15 million °C (27 million °F).

Prominences and filaments, flamelike plumes of hot plasma, stretch across the surface of the Sun

the formation of the Solar System

About 4.6 billion years ago, a giant cloud of interstellar gas and dust started to collapse in on itself, perhaps triggered by the shockwaves from a star exploding nearby. As the cloud collapsed, it started to spin and heat up, forming a flat, swirling disc of gas and dust. Once the heat and pressure at the core of the nebula were high enough, hydrogen atoms started to fuse into helium atoms, releasing a huge amount of energy. A new star, our Sun, was born at the heart of the rotating disc. Gravity drew the remaining materials in the disc together, creating asteroids, comets, planets, and moons.

HOW THE PLANETS FORMED

The material in the disc around the young Sun included dust, gas, and in the cold outer reaches, fragments of ice. This material collided and sometimes stuck together in a process called accretion. Fragments of dust clumped together to form pebbles, pebbles came together to form rocks, and rocks grew into planetary building blocks called planetesimals. Some of these objects grew large enough for gravity to shape them into spheres, and they became planets or moons.

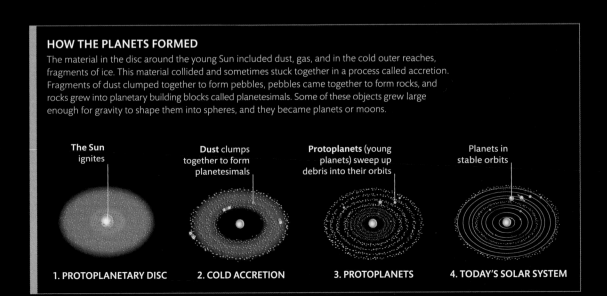

The Sun ignites

Dust clumps together to form planetesimals

Protoplanets (young planets) sweep up debris into their orbits

Planets in stable orbits

1. PROTOPLANETARY DISC
2. COLD ACCRETION
3. PROTOPLANETS
4. TODAY'S SOLAR SYSTEM

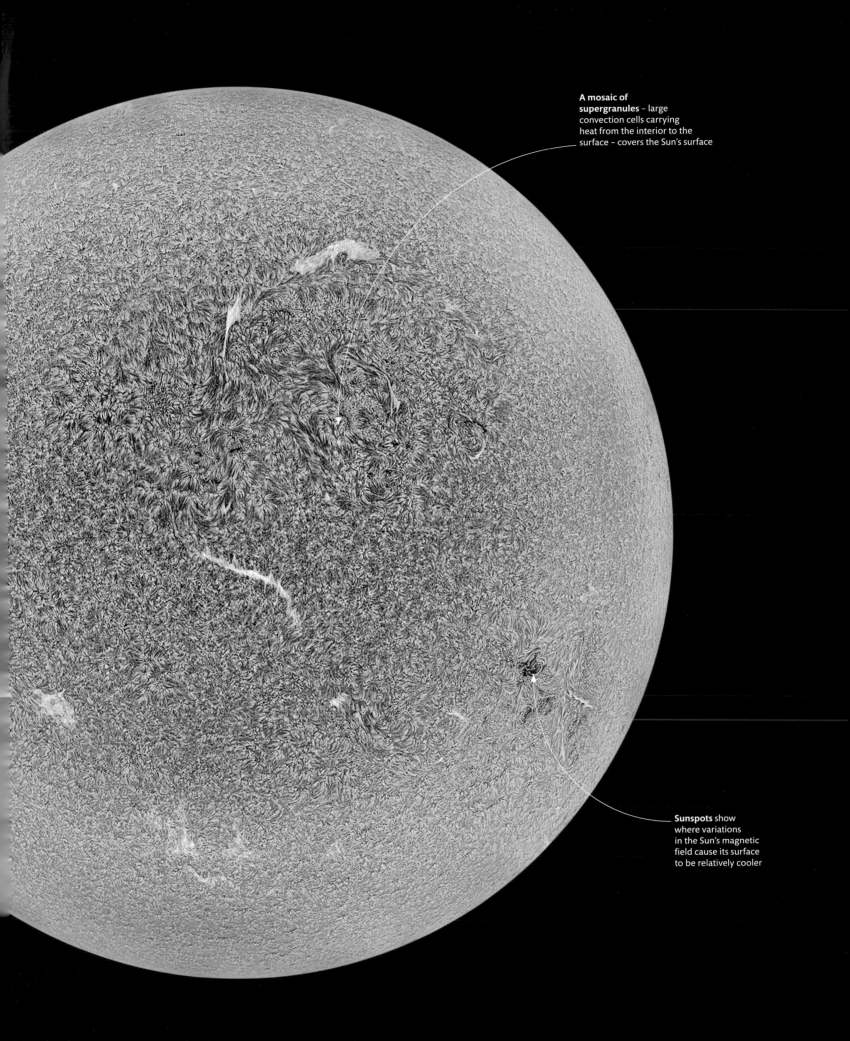

A mosaic of supergranules – large convection cells carrying heat from the interior to the surface – covers the Sun's surface

Sunspots show where variations in the Sun's magnetic field cause its surface to be relatively cooler

Chondrite micrometeorite is less than 2 mm (¹⁄₁₆ in) across

Micrometeorite
Shown here is a scanning electron micrograph of a tiny meteorite that was found on a beach on the east coast of the US. This specimen belongs to a class of meteorites called chondrites that are made up of the same primitive material that originally formed the rocky planets.

meteorites

Earth sweeps up about 70,000 tonnes of extraterrestrial material each year. While most of it consists of microscopic dust particles, a significant proportion of larger objects fall to the surface to become meteorites. Today, most meteorites come from the Asteroid Belt between Mars and Jupiter, where collisions between the asteroids can send fragments of rocks towards Earth. Among these are meteorites containing carbon and even complex organic molecules, which may have played a role in the origin of life. Like all planets, Earth was showered with objects when the Solar System was young. The Moon bears the scars of this bombardment, but the continual recycling of Earth's crust through plate tectonics has erased all but the most recent impact craters from our planet's surface.

Fragments of rock suggest this fragment formed at the boundary between the metallic core and the silicate mantle of the asteroid

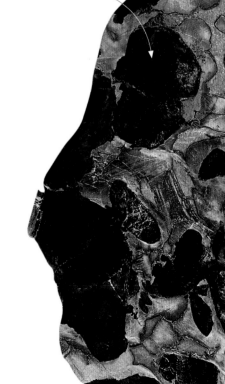

IMPACT CRATERS

If an asteroid or meteoroid is large enough and travelling fast enough, it leaves a circular crater when it hits the ground and explodes. The kinetic energy of the speeding rock is instantly converted into heat that may be sufficient to vapourize the meteorite and some of the surface rock. A part of the meteorite may survive as melted rock or broken fragments (breccia), and some of the debris thrown out settles into a halo of ejecta. The rock layers beneath the crater may be fractured, uplifted, or overturned.

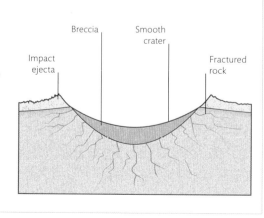

Impact ejecta | Breccia | Smooth crater | Fractured rock

Core remnant
These cut and polished sections of stony-iron meteorites found in Seymchan, Russia, show crystals of iron and nickel that were formed in the molten core of a planetesimal or a differentiated asteroid – one that has segmented into a core, mantle, and crust. Some meteorites are rocky, with a few showing signs of melting in their parent bodies. Others have a chemical composition similar to the Sun's and are thought to be representative of the solar nebula from which the planets formed.

Olivine (magnesium iron silicate) crystals embedded in iron-nickel metal

Crystals of nickel and iron formed in the hot core of an asteroid

The distinctive texture of iron-nickel alloy crystals, which appears when a cross section of an iron meteorite is cut, polished, and etched with weak acid, is called the Widmanstätten pattern

Erosion and resurfacing have obliterated most of Earth's many impact craters, leaving only the youngest. Old, stable rocks and lack of vegetation make central Australia a good place to find them. One of the few remnants of Earth's bombardment by comets and asteroids, the Tnorala impact feature, also known as Gosse's Bluff, is located about 200 km (125 miles) west of Alice Springs in Australia's Northern Territory.

spotlight Tnorala

Tnorala was created in the early Cretaceous period, 142 million years ago, when an asteroid or comet 600 m (2,000 ft) across, hurtling at 40 km/s (25 miles per second), slammed into the plains of central Australia, leaving behind a 22-km- (14-mile-) wide puncture in Earth's crust. Today, erosion has removed almost all traces of the outer rim of this impact crater, but the central uplift remains in the form of a circular ring of sandstone hills about 5 km (3 miles) across. Although eroded flat, the crater's rim is an outer circle of darker rock seen in satellite images. The land surface of Tnorala now lies about 2 km (1.25 miles) lower than it was at the time of impact. The collision left evidence including deformed and fragmented rocks, impacted and melted quartz grains, and most compellingly, shatter cones in the rock, left by shockwaves radiating from the impact.

Tnorala is a sacred site for the Western Aranda Aboriginal people, whose folklore also attributes its creation to an object falling from the sky. In their story, a baby in a basket falls to Earth, dropped by a group of celestial women dancing along the Milky Way. The rocks were pushed up when the basket landed.

A "horsetail" pattern of thin grooves is left behind in rocks subjected to a high-pressure shockwave

Shatter cone

Sandstone ring
Composed of a circle of sandstone hills, Tnorala rises 180 m (600 ft) above the surrounding plain. It is one of Australia's most studied impact craters. Its impact origin was first proposed in the 1960s, based largely on the abundance of shatter cones found in the area.

The Moon's surface is pitted with scars from ancient meteoroid, asteroid, and comet impacts

Double planet

The astronauts on the Apollo 17 mission were the last people to see Earth from the Moon. They took this photo of a crescent Earth rising above the lunar surface in December 1972. The Moon is so large relative to Earth that planetary scientists consider them to be a double planet. However, they are strikingly different bodies: one is geologically active and teeming with life and the other is a barren relic from the early Solar System.

Earth is almost four times the size of the Moon – large enough to retain a thick atmosphere, protected from the solar wind by a powerful magnetic field

Moon rock

Rock samples from the Moon collected by astronauts on NASA's Apollo programme between 1969 and 1972 reveal that the Moon's surface is similar in composition to Earth's mantle. Up to 4.4 billion years old, these specimens help date Earth's early history and the Moon's origins.

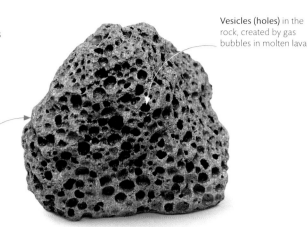

Vesicles (holes) in the rock, created by gas bubbles in molten lava

This 3.5-billion-year-old rock sample brought back by astronauts aboard the Apollo 15 mission is similar to the basalt rocks found around Hawaii, US

the Moon

It is thought that less than 100 million years after Earth formed, it was struck by a smaller planet about the size of Mars. This cataclysmic impact melted the smaller planet and a significant part of Earth's crust, and splashed the hot debris into space, where it coalesced and solidified to form Earth's only natural satellite. The young Moon was bombarded by comets and asteroids, which have left craters all over its surface. Some of the largest impact basins have been flooded by lava, creating the dark "seas" seen from Earth. The far side of the Moon is more mountainous. The Moon is the Solar System's largest satellite relative to the size of its parent planet, and has had a correspondingly large influence on the evolution of Earth and its life forms. As well as driving the ocean tides, the Moon has stabilized Earth's rotation rate, axial tilt, and climate. It may also have enhanced Earth's protective magnetic field. It looms large in our night sky and was an early target for space exploration.

THE MOON'S ORBIT

The gravitational tug of the Moon is the main cause of tides in Earth's oceans, but the Moon also pulls at our planet's rocks, and vice-versa. This exchange of gravitational energy has slowed Earth's rotational period from about 5 hours when the Moon was young to 24 hours today. It has also increased the diameter of the Moon's orbit from 128,000 km (80,000 miles) to 383,000 km (238,000 miles) and tidally locked the Moon – which means it has the same rotation period as Earth – so only one side, called the near side, is visible from Earth.

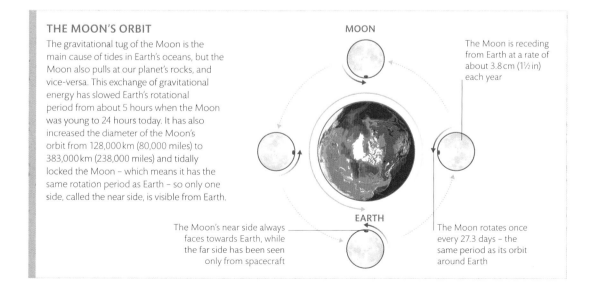

MOON

The Moon is receding from Earth at a rate of about 3.8 cm (1½ in) each year

EARTH

The Moon's near side always faces towards Earth, while the far side has been seen only from spacecraft

The Moon rotates once every 27.3 days – the same period as its orbit around Earth

our place in the Solar System

Once the planets acquired enough mass, their own gravity started to compress them, transforming gravitational energy into heat energy. This, along with the internal heat from radioactive decay, was enough to melt their interiors, allowing heavier elements to sink towards the core and lighter elements to float towards the surface in a process known as planetary differentiation. This left Earth with a solid crust and a large, partly liquid iron core capable of generating a magnetic field powerful enough to prevent the stream of charged particles known as the solar wind from stripping away its atmosphere. Earth is uniquely structured and positioned to allow life to thrive.

Volcanic eruptions and frequent meteorite impacts kept Earth's surface molten until a crust formed as the planet cooled

Hell on Earth
The surface of early Earth would have been a hellish place, caught between protoplanetary debris raining down from above and heat rising up from the layers below.

SOLAR SYSTEM ORBITS

The Solar System is a family of bodies held in position by the gravitational pull of the Sun. The bodies are grouped into rocky inner planets (Mercury, Venus, Earth, and Mars), a belt of tiny asteroids, the gas and ice giants (Jupiter, Saturn, Neptune, and Uranus), the Kuiper Belt of distant icy bodies that includes the dwarf planet Pluto, and the more distant Oort Cloud of comets. The astronomical unit (AU) is used to compare distances in the Solar System: 1 AU is the average distance between Earth and the Sun. Earth is the only planet within the Solar System's habitable zone (shown in green below) – it is neither too hot nor too cold, so liquid water is stable on its surface.

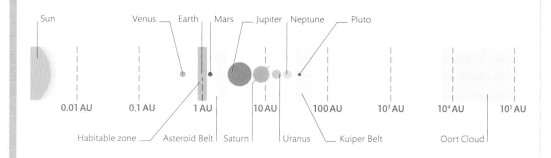

Ice-filled crater

Mars has enough water to flood its surface to a depth of at least 30 m (100 ft). However, liquid water cannot exist on Mars due to the low atmospheric pressure and the extremely low temperature. Mars lies beyond the outer edge of the Solar System's habitable zone. Its water is trapped as ice, mostly below the surface, but sometimes also on the floors of craters, such as Korolev Crater, near the poles.

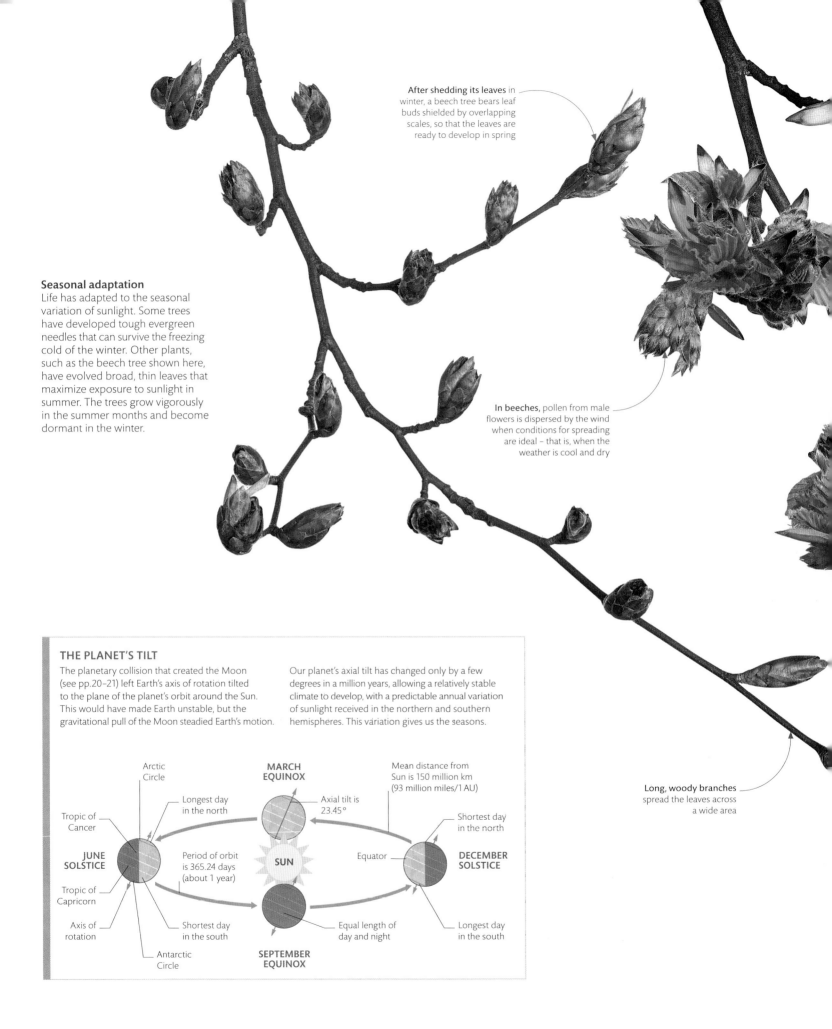

After shedding its leaves in winter, a beech tree bears leaf buds shielded by overlapping scales, so that the leaves are ready to develop in spring

Seasonal adaptation

Life has adapted to the seasonal variation of sunlight. Some trees have developed tough evergreen needles that can survive the freezing cold of the winter. Other plants, such as the beech tree shown here, have evolved broad, thin leaves that maximize exposure to sunlight in summer. The trees grow vigorously in the summer months and become dormant in the winter.

In beeches, pollen from male flowers is dispersed by the wind when conditions for spreading are ideal – that is, when the weather is cool and dry

Long, woody branches spread the leaves across a wide area

THE PLANET'S TILT

The planetary collision that created the Moon (see pp.20–21) left Earth's axis of rotation tilted to the plane of the planet's orbit around the Sun. This would have made Earth unstable, but the gravitational pull of the Moon steadied Earth's motion.

Our planet's axial tilt has changed only by a few degrees in a million years, allowing a relatively stable climate to develop, with a predictable annual variation of sunlight received in the northern and southern hemispheres. This variation gives us the seasons.

Arctic Circle

MARCH EQUINOX

Mean distance from Sun is 150 million km (93 million miles/1 AU)

Longest day in the north

Axial tilt is 23.45°

Tropic of Cancer

Shortest day in the north

JUNE SOLSTICE

Period of orbit is 365.24 days (about 1 year)

SUN

Equator

DECEMBER SOLSTICE

Tropic of Capricorn

Axis of rotation

Shortest day in the south

Equal length of day and night

Longest day in the south

Antarctic Circle

SEPTEMBER EQUINOX

days and seasons

Earth rotates on its axis once every 23 hours and 56 minutes. It also revolves in its orbit, and so it takes a little longer for the Sun to return to the same point in the sky, giving us the length of one day, which we split into 24 hours. At the equator, we experience 12 hours of daylight and 12 hours as night, but as we move north or south of the equator, the tilt of Earth's rotational axis causes the days to be longer in summer and the nights to be longer in winter. This daily and annual variation in energy received from the Sun is what drives Earth's seasons.

Brown bud scales, which protected the buds from cold in winter, are shed in spring, as the buds burst open

New leaves burst out from their buds in the spring

The green chlorophyll in leaves helps combine sunlight with water and nutrients to fuel the growth of the tree

Midnight Sun

As we move north of the Arctic Circle or south of the Antarctic Circle (66.5° North and South), there is at least one day in summer when the Sun does not set, and one day in winter when it does not rise. The multiple-exposure image here shows the Sun above the Arctic Circle dipping towards the horizon close to midnight. It does not quite set and instead, rises immediately after midnight. The North and South poles experience 6 months of daylight followed by 6 months without seeing the Sun.

Foucault's pendulum in Paris
This replica of Foucault's original
apparatus is installed in the
Paris Panthéon. Using a system
originally devised by Vincenzo
Viviani, Foucault used a wire
67 m (220 ft) in length to suspend
a 28 kg (62 lb) lead weight from
the dome of the building.

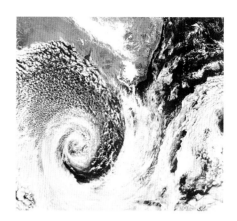

Spiral of weather
This satellite view of a low-pressure system over Australia shows the influence of the Coriolis effect. Resulting from Earth's rotation, it causes storms and cyclones to spin: anticlockwise in the northern hemisphere and clockwise (as here) in the southern hemisphere.

history of Earth science

Earth's rotation

Although astronomers in India in the 5th century BCE, and in the Islamic world in the 10th century, believed that Earth rotates once a day, the accepted view in 16th-century Europe was that the heavens revolved around a fixed Earth. In 1543 Nicolaus Copernicus proposed a rotating Earth as part of his heliocentric model, but a century later critics were still rejecting his theory for lack of physical evidence of Earth's spin.

The search for proof of Earth's rotation centred on experiments to test whether a moving object deflects slightly sideways – a phenomenon now known as the Coriolis effect. It was not until the late 18th century that a measurable effect was confirmed, when weights were observed to deflect by a few centimetres when dropped from towers more than 150 m (500 ft) tall in height.

Much clearer evidence for Earth's rotation was provided by French physicist Jean Bernard Léon Foucault in 1851. He built a long pendulum with a special mount that allowed it to rotate freely.

If Earth is not rotating, the pendulum would be expected to swing in a fixed plane, conforming to the principle of conservation of angular momentum. However, if Earth is rotating, angular momentum would be conserved by the plane of the swing rotating (precessing) relative to Earth's surface.

For a pendulum at the north or south pole, the swing plane would complete a full 360-degree rotation in 24 hours. At lower latitudes the effect is less, so the precession is slower, falling to zero at the equator. At the latitude of Paris, where Foucault conducted his experiment, his pendulum took almost 32 hours to complete one rotation, precessing clockwise at about 11 degrees per hour. Demonstrations of his experiment in different locations across Europe and America attracted huge public interest.

Although small on a human scale, the effects of rotation are highly significant on a planetary scale, giving rise to the circulation patterns of Earth's atmosphere and oceans.

> " You are invited to come to see the Earth turn, tomorrow, from three to five, at Meridian Hall of the Paris Observatory. "

LÉON FOUCAULT, INVITATION CARD, 3 FEBRUARY 1851

RADIOMETRIC DATING

Certain Earth materials, including the ancient zircon below, can be aged by radiometric dating. This uses the slow but clocklike decay of radioactive isotopes, such as uranium-235 to lead, to provide an absolute scale against which we can measure geological time. The ratio of uranium-235 to lead in a rock therefore tells us its age. The isotopes of other elements can tell us about conditions when some rocks were formed. For instance, the ratio of oxygen-18 to oxygen-16 tells us about the presence of liquid water, and the ratio of carbon-12 to carbon-13 about the presence of life.

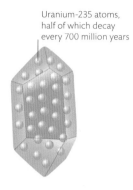

Uranium-235 atoms, half of which decay every 700 million years

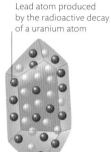

Lead atom produced by the radioactive decay of a uranium atom

One-quarter of the original uranium atoms remain

One-eighth of the uranium atoms remain

| NEW ROCK | 700 MILLION YEARS | 1,400 MILLION YEARS | 2,100 MILLION YEARS |

Zircon crystal
is just 0.4mm tall and is barely visible to the naked eye

Parallel crystal
growth lines

Blue colour in
this image results from bombardment by electrons under the microscope

Extra-terrestrial water

Some of Earth's water will have arrived on comets that frequently impacted the young planet. Comets are composed of about half ice and half dust. They lose some of their ice, which is about 80 per cent water, through spectacular geyser-like eruptions as they warm up on entering the inner solar system. The Rosetta spacecraft captured this in close-up as it approached Comet 67P/Churyumov-Gerasimenko in 2014.

Traces of the first ocean

The oldest materials yet found on Earth are crystals of the mineral zircon, formed 4.4 billion years ago. They contain oxygen isotopes that indicate the presence of liquid water at that time. Zircon is as tough as diamond and a long-lived geological record keeper.

origin of the oceans

As Earth settled into layers in the process of differentiation (see p.22), volatile materials were expelled from its interior by intense volcanic activity, producing an atmosphere of nitrogen, carbon dioxide, and water vapour. The surface quickly cooled to form a solid crust, allowing water vapour to condense and pool, forming the first ocean. There is evidence from ancient crystals of zircon found in the Jack Hills of Western Australia that surface water was present at least 4.4 billion years ago, only 160 million years after the Earth formed. Some meteorites contain 15–20 per cent water, and early Earth is thought to have had a similar composition, providing enough water for the first ocean.

Old at heart
The oldest rocks are found in the stable centres of some continents – remnants of Earth's early cratons. This metamorphic rock from an island in the Acasta River in Canada is about 4 billion years old.

During this rock's long history, bands were produced by metamorphosis under high heat and pressure

the continents form

During the process of differentiation (see p.22) Earth's interior shaped itself into distinct layers. A metallic core formed beneath a silicate-rich mantle and a lightweight crust. This primordial crust was uniform, but it too began slowly differentiating, into two types – oceanic and continental crust. The crust, fused to the uppermost layer of the mantle, floated on hot, mobile material beneath. The heat from below caused the outer layers to move and fracture into tectonic plates. Some of these plates were dragged down into the mantle, while new crust formed where the plates were pushed apart. This repeated recycling of the crust caused the concentration of lighter elements in some locations, creating the thick, buoyant continental crust that now covers about 40 per cent of Earth's surface.

Waterfall of lava
Lava pours into the Pacific Ocean, vapourizing the water at Kamokuna lava delta on Hawaii's Big Island. The island is gradually growing as hot, runny lava flows from the Kilauea volcano into the sea, adding material to the island's broad flanks. This process, continuing today in Hawaii, was also part of the process of early continent building.

HOW THE FIRST CONTINENTS FORMED

About 4 billion years ago, Earth's plates began to move, and some of the primordial crust was dragged down into the mantle. The crust melted and released water, causing the mantle to melt, producing magma rich in light elements that rose up to form volcanic islands. Plate motion pushed the islands together to form cratons – masses of light rock. Weathering and sedimentation further concentrated the lighter materials on the cratons. When plate motion brought cratons together, they formed larger masses of continental crust.

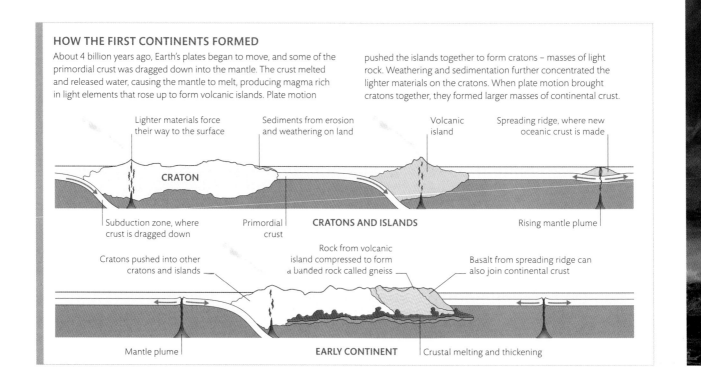

Lighter materials force their way to the surface

Sediments from erosion and weathering on land

Volcanic island

Spreading ridge, where new oceanic crust is made

CRATON

Subduction zone, where crust is dragged down

Primordial crust

CRATONS AND ISLANDS

Rising mantle plume

Cratons pushed into other cratons and islands

Rock from volcanic island compressed to form a banded rock called gneiss

Basalt from spreading ridge can also join continental crust

Mantle plume

EARLY CONTINENT

Crustal melting and thickening

the ages of Earth

If Earth's history were to be compressed into one day, the dinosaurs would appear for about an hour at 10:40pm, and our human ancestors would not arrive until two minutes to midnight. Instead of hours, minutes, and seconds, geological time is split into eons, eras, and periods. These intervals do not mark equal slices of time, but what can be interpreted in the rock record. The Hadean eon (derived from the Greek word for "beneath the Earth") is the first; unfortunately, there are almost no surviving rocks from this time. Archaean (ancient) rocks contain sparse evidence of bacteria, but as rocks get younger, through the Proterozoic (earlier life) to the Phanerozoic eon (visible life), they begin to contain more and more evidence of the development of continents, oceans, their environments, and the evolution of living organisms.

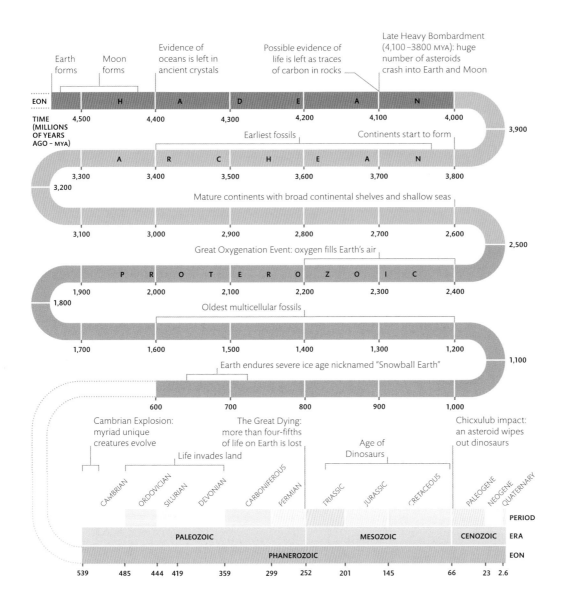

Earth forms

Moon forms

Evidence of oceans is left in ancient crystals

Possible evidence of life is left as traces of carbon in rocks

Late Heavy Bombardment (4,100–3800 MYA): huge number of asteroids crash into Earth and Moon

EON H A D E A N

TIME (MILLIONS OF YEARS AGO – MYA) 4,500 — 4,400 — 4,300 — 4,200 — 4,100 — 4,000 — 3,900

Earliest fossils

Continents start to form

A R C H E A N

3,300 — 3,400 — 3,500 — 3,600 — 3,700 — 3,800

3,200

Mature continents with broad continental shelves and shallow seas

3,100 — 3,000 — 2,900 — 2,800 — 2,700 — 2,600 — 2,500

Great Oxygenation Event: oxygen fills Earth's air

P R O T E R O Z O I C

1,900 — 2,000 — 2,100 — 2,200 — 2,300 — 2,400

1,800

Oldest multicellular fossils

1,700 — 1,600 — 1,500 — 1,400 — 1,300 — 1,200 — 1,100

Earth endures severe ice age nicknamed "Snowball Earth"

600 — 700 — 800 — 900 — 1,000

Cambrian Explosion: myriad unique creatures evolve

The Great Dying: more than four-fifths of life on Earth is lost

Life invades land

Age of Dinosaurs

Chicxulub impact: an asteroid wipes out dinosaurs

CAMBRIAN · ORDOVICIAN · SILURIAN · DEVONIAN · CARBONIFEROUS · PERMIAN · TRIASSIC · JURASSIC · CRETACEOUS · PALEOGENE · NEOGENE · QUATERNARY

PERIOD

| PALEOZOIC | MESOZOIC | CENOZOIC | **ERA** |

PHANEROZOIC **EON**

539 — 485 — 444 — 419 — 359 — 299 — 252 — 201 — 145 — 66 — 23 — 2.6

The story of our planet
The geological timescale provides a chronology of major events in Earth's history since it formed 4.6 billion years ago. In this chart, the most recent eon, the Phanerozoic, is divided into eras, which are further subdivided into periods. The timescale is marked by the appearance and extinction of new life forms.

Rock record
In some places, Earth's rock layers can be read like the pages of a history book. The layers of shale and clay at Bentonite Hills, Utah, US, resulted from the deposition of mud, silt, fine sand, and volcanic ash in swamps and lakes during the Jurassic period. By studying rock layers like these, we can learn what life on Earth was like 145 million years ago.

Earth's materials

Earth's thin outer crust is made of a great diversity of rocks that have developed as a result of billions of years of geological activity. The natural building blocks of rocks – and all the solid bodies of the Universe – are minerals. The crust is overlain and partly concealed by an even thinner layer of soil and vegetation, and by bodies of liquid and frozen water.

Marcasite crystals

The mineral marcasite is a polymorph of pyrite, meaning it has the same chemical formula as pyrite but a slightly different crystal structure. Unlike pyrite, marcasite tarnishes and breaks down quickly when exposed to air.

Spear-shaped marcasite crystals have a different shape to pyrite crystals and belong to a different crystal system

Pyrite crystals

Pyrite is found in almost every environment on Earth. Its crystals can be shaped as cubes, octahedrons, or 12-sided pyritohedrons. This cubic crystal is from Guangxi Province in China. Pyrite's name is derived from the Greek word for "fire", and it gives off sparks when struck hard with a hammer.

Cubes are the most symmetrical of all crystal shapes, with three equal axes of symmetry at right angles and four-fold symmetry (it looks the same at four points during a complete rotation)

crystal structure

Minerals are inorganic solids that occur naturally and are made up of chemical elements whose atoms are typically arranged in a systematic internal pattern. When free to grow in an open space, minerals develop regular geometric shapes called crystals, which can often be seen with the naked eye. These are a reflection of a mineral's internal crystal structure, the regular arrangement in three-dimensional units of its atoms or ions – ions are atoms or molecules that have an electric charge. One way of describing the symmetry of a crystal is through axes of symmetry: imaginary lines drawn through a crystal's centre from the middle of opposite faces. Crystals are divided into seven groups, or crystal systems, each with a characteristic shape and symmetry. Most crystals do not achieve a perfect crystal shape because they form in restricted spaces in nature.

With its yellow colour and metallic lustre, pyrite is sometimes mistaken for gold, earning it the nickname "fool's gold"

CRYSTAL TWINNING

Two or more crystals of the same mineral sometimes grow symmetrically. These are called crystal twins. There are two main types: contact and penetration twins. In contact twins, there is a distinct boundary between the crystals, which appear as mirror images. In penetration twins, individual crystals appear to grow through each other. The world's most abundant minerals, feldspars, are often twinned. Quartz and spinel form contact twins, while orthoclase feldspar, pyrite, and fluorite form penetration twins. The type of twinning can help to identify a mineral.

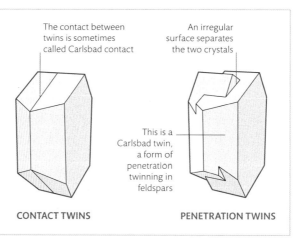

The contact between twins is sometimes called Carlsbad contact

An irregular surface separates the two crystals

This is a Carlsbad twin, a form of penetration twinning in feldspars

CONTACT TWINS

PENETRATION TWINS

Striations, or linear stripes or marks, on pyrite's cubic crystal faces are usually caused by two different crystal forms developing at the same time

Forms and faces

The term "form" describes a set of identical crystal faces that have symmetrical planes. Forms are referred to as "closed" or "open". In closed forms, such as cubes or octahedra, the faces are identical. In contrast, open forms have faces of different shapes and sizes. For example, a specimen that has mainly parallel prism faces (see right) has to terminate in a different type of face (here, a pyramid) to enclose the space. If a crystal consists of more than one set of faces, the habit is named for the faces that predominate (here, prismatic).

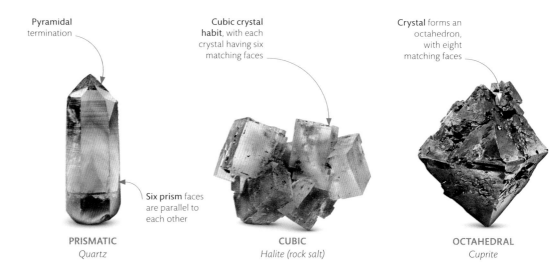

Pyramidal termination

Six prism faces are parallel to each other

PRISMATIC
Quartz

Cubic crystal habit, with each crystal having six matching faces

CUBIC
Halite (rock salt)

Crystal forms an octahedron, with eight matching faces

OCTAHEDRAL
Cuprite

Appearance

Some habits are named for their general appearance, rather than for the crystal's forms and faces. This is the case with aggregates, which are crystals that grow together in a group rather than as separate, individual crystals, generally because they are imperfectly developed. In some aggregates, the crystals are microscopic. There are many different terms used to describe aggregates, which are highly variable in size and habit.

Crystal is solid, with no obvious structure; the crystals cannot be seen individually

MASSIVE
Dumortierite

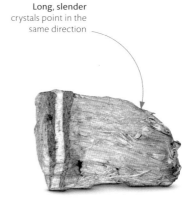

Long, slender crystals point in the same direction

FIBROUS
Tremolite

Long, thin crystals spread out from a central point

RADIATING
Pyrophyllite

Crystals are elongated and flattened, with a curved edge, like a knife blade

BLADED
Kyanite

Mass of slender, long crystals resembling needles may radiate from a central point

NEEDLE-LIKE
Mesolite

Clusters of crystals form in slender, divergent, fern-like branches

DENDRITIC
Copper

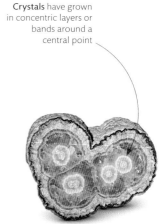

Crystals have grown in concentric layers or bands around a central point

CONCENTRIC
Rhodochrosite

crystal habits

Habit refers to a crystal's external shape, including the forms and faces of individual crystals. The description can refer to a single, well-formed crystal or to a cluster of intergrown crystals, known as an aggregate. In nature, perfectly formed crystals are rare, because their development is limited by the size and shape of the cavities in which they grow, as well as by gravity. It is possible for a mineral to have a variety of habits depending on the conditions in which it forms.

Crystals are globular and form in clusters, like a bunch of grapes

BOTRYOIDAL
Malachite

Flat, slab-like crystals form rectangular or square parallel faces

TABULAR
Torbernite

Prismatic crystal forms a pyramidal termination with six faces

This crystal has grown into a long, hexagonal column or prism

Vitreous lustre typical of all beryl crystals

Aquamarine
The blue variety of beryl, aquamarine forms deep inside Earth and is often associated with granite intrusions (see pp.134–35). Its crystals form six-sided prisms, with its faces running parallel to a central axis. This specimen is from the Jos Plateau in Nigeria.

Silvery white colour is typical of freshly produced bismuth; over time it can develop an iridescent sheen

Brilliant, shiny lustre gives diamond a sparkly look

ADAMANTINE
Diamond

Surface of mineral is rough and non-reflective

DULL
Hematite

Unreflective surface resembles dried mud, soil, or clay

EARTHY
Bentonite

Microscopic irregularities on the surface give the mineral a greasy look

GREASY
Chrysocolla

Slight shimmer given off by the mineral's parallel fibres

SILKY
Ulexite

Crystal prisms gleam like glass

VITREOUS
Quartz

Shiny surfaces
resemble aluminium
foil due to their lustre

Flat surfaces
catch the light

reflecting light

A mineral is often described or identified by its lustre – that is, its reflective qualities and the extent of its sheen. Minerals can be broadly referred to as metallic, submetallic, or non-metallic. Those described as having metallic lustre are reflective and opaque to light, like polished metal. Submetallic minerals are duller and less reflective, and minerals with non-metallic lustre are usually lighter in colour, with many showing some degree of transparency or translucency. There are several types of non-metallic lustre, some of which are shown opposite.

Metallic lustre
Bismuth, with its brilliant, highly reflective surfaces that gleam like metal, is described as having metallic lustre. As with all similar minerals, the "polished metal" effect is produced when light striking the surface stimulates electrons, causing them to vibrate and emit a diffuse light.

REFRACTION AND INTERNAL REFLECTION

The extent of a crystal's transparency depends on what happens to light rays as they pass through the crystal. If the incident ray (incoming light) is at right angles to the crystal-air interface, all the light goes straight through the crystal (1). More often, the ray is at a different angle, and the light is partly refracted and partly reflected as it slows down and changes direction (2). As the angle of incidence increases, the proportion of reflected light also increases. The critical angle (3) is the angle of incidence above which total internal reflection occurs. When the angle of incidence is greater than the critical angle, all light is reflected (4).

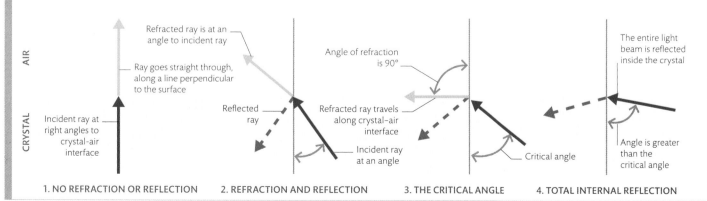

Refracted ray is at an angle to incident ray

Ray goes straight through, along a line perpendicular to the surface

Incident ray at right angles to crystal-air interface

Reflected ray

Angle of refraction is 90°

Refracted ray travels along crystal-air interface

Incident ray at an angle

The entire light beam is reflected inside the crystal

Critical angle

Angle is greater than the critical angle

AIR

CRYSTAL

1. NO REFRACTION OR REFLECTION   2. REFRACTION AND REFLECTION   3. THE CRITICAL ANGLE   4. TOTAL INTERNAL REFLECTION

Pearly green
lustre

Talc
The softest of all minerals, talc has a value of 1 on the Mohs Hardness Scale and can be easily scratched by a fingernail. Cosmetics, including talcum powder, are made from ground-down talc.

Rough diamond
This 424-carat diamond was found in South Africa in 2019. Diamonds are the hardest naturally occurring substances; as such, they can scratch any other mineral, but only a diamond can scratch a diamond. These precious gems are pure carbon and form about 160 km (100 miles) below Earth's surface, under great pressure and intense heat. The diamonds we see at the surface have been carried upwards by very deep-seated volcanic eruptions.

from hard to soft

Testing hardness is a very useful way to identify an unknown mineral. The hardness of a mineral refers to the extent of its resistance to scratching or abrasions, rather than its toughness or strength; very hard minerals can actually be quite brittle and break easily. A scratch mark on a mineral marks the place where atoms have been removed from its surface. The strength of the bonds that hold the atoms together can influence a mineral's hardness. For example, graphite is a relatively soft mineral, with weak bonds between its atoms, whereas diamonds, which are the hardest minerals, have strong atomic bonds. All minerals can be assigned a hardness value using one of several testing methods, including Mohs scale and the Knoop scale.

Clear, colourless
specimen is made of
pure carbon, with no
trace elements

ASSESSING HARDNESS

The most widely used hardness test for minerals is called Mohs scale. This measures hardness relative to 10 standard minerals of increasing hardness, from 1 (as soft as talc) through to 10 (as hard as a diamond). To test a mineral's hardness, it can be scratched using a known mineral (a harder mineral will scratch a softer one, but not vice versa) or a common object of known hardness. For example, a fingernail, copper coin, and steel file will scratch minerals that are less than 2.5, 3.5, and 6.5 respectively. For the Knoop test, a load is applied to the surface of the sample using an indenting tool. The hardness value is based on the ratio of the applied load to the area of indentation.

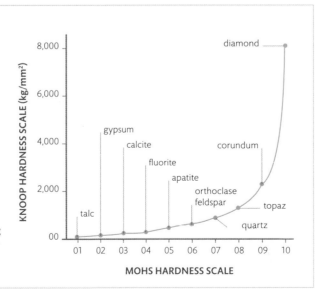

Irregular shape
with rounded edges

native elements

Most minerals are made up of a combination of chemical elements in compounds. However, some elements, known as native elements, occur in nature in a relatively pure form. Common native elements are divided into three groups: metals (gold, silver, platinum, copper, and iron), semi-metals (arsenic and bismuth), and non-metals (sulfur and carbon). Other native elements occur more rarely. Native elements occur in various rock types, and are often of economic value. The minerals gold, silver, platinum, osmium, and iridium are the main sources for the elements with the same name.

Native copper
Probably the first metal to be worked by humans, copper is a bright, reddish gold metal that tarnishes to dull brown, and often forms green or black crusts when exposed to oxygen. Traditionally used to make coins along with silver and gold, copper conducts heat and electricity well, and is used to make many kinds of electrical equipment.

Wiry, branching
tendrils form
a mass

Rounded nugget
shape is typical
of platinum

SILVER

PLATINUM

Flat cleavage plane
along which
graphite breaks

Well-formed,
bright yellow
crystals

GRAPHITE

SULFUR

NATURAL FORMS OF CARBON

Carbon is one of the most abundant chemical elements on Earth and exists in many forms. Diamond and graphite are both crystalline: diamond develops in high-temperature, high-pressure conditions and graphite in high temperatures. Amorphous carbon (which includes soot and coal) is non-crystalline and forms through incomplete combustion. Carbon is rare among elements in producing long, strongly bonded chains of atoms called polymers.

Each carbon atom
is attached to four
other carbon atoms

Carbon atoms are arranged
in layers of hexagons

Carbon atoms
are disordered

DIAMOND

GRAPHITE

AMORPHOUS CARBON

Warm, reddish brown
background colour

Green patches of tarnish caused by oxidation are visible on the surface

Distorted, branched appearance is typical for copper

Metal compound
Most metals combine with non-metallic chemical elements or with semi-metals to form a great variety of minerals. For example, tetrahedrite (shown here) is a compound of the metal copper with the chemical element sulfur and the semi-metal antimony. Semi-metals have properties that are intermediate between those of metals and non-metals.

The triangular pyramid-shaped (tetrahedral) crystals give the mineral its name

Surface is shiny and reflective

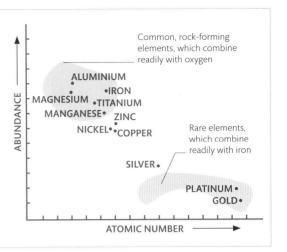

ABUNDANCE OF METALS IN EARTH'S CRUST

This diagram shows that the most abundant metals in Earth's crust are common, rock-forming elements that combine with oxygen to form silicate minerals such as quartz and feldspar. These elements include iron, aluminium, magnesium, titanium, and manganese. The rarest metals are elements that are depleted in the crust because they mainly reside deeper within Earth's interior. These rare metals include platinum and gold.

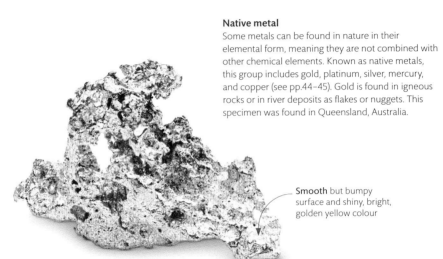

Native metal
Some metals can be found in nature in their elemental form, meaning they are not combined with other chemical elements. Known as native metals, this group includes gold, platinum, silver, mercury, and copper (see pp.44–45). Gold is found in igneous rocks or in river deposits as flakes or nuggets. This specimen was found in Queensland, Australia.

Smooth but bumpy surface and shiny, bright, golden yellow colour

Uneven fracture of the crystals

Earth's metals

Metals occur in various forms in Earth's crust. They are typically defined as solid, shiny, and opaque, are usually crystalline, and comprise the main group of chemical elements in the periodic table. Not all metals feature in the table, however, as some form compounds with other metals or non-metals – for example, steel is not included since it is a mix of iron, nickel, and chromium. Ores are rocks that contain valuable metals or metal compounds in amounts that are large enough for their extraction to be commercially worthwhile. Metals conduct electricity and heat well, and are both malleable and resilient, making them sought after for many uses.

Concentric layers, each with a different chemical composition

Tiny, rounded crystals form in clusters, resembling a bunch of grapes

Deep orange colour is due to the presence of iron oxide

Rounded crusts of tiny whitish crystals have a pink tinge

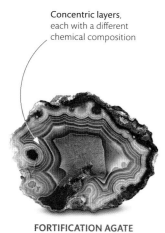

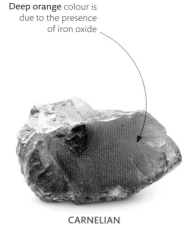

FORTIFICATION AGATE

GRAPE AGATE

CARNELIAN

PINK CHALCEDONY

Microcrystalline quartz
In some varieties of quartz, described as microcrystalline (or cryptocrystalline), the crystals are too small to be seen with the naked eye and are visible only under high magnification. Microcrystalline quartz typically forms at low temperatures in volcanic settings.

Rich purple, pyramid-shaped crystals

Pyramid-shaped, orange or yellow-brown crystals

Colourless, transparent, long prismatic crystals

Translucent pink crystals of varying lengths

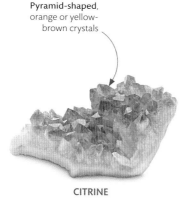

AMETHYST

CITRINE

ROCK CRYSTAL

ROSE QUARTZ

Macrocrystalline quartz
Quartz varieties with crystals that are visible to the naked eye are described as macrocrystalline. The crystals are six-sided prisms and pyramids, often perfectly formed, although less well-developed crystals form imperfect shapes. Aggregates or crusts of macrocrystalline quartz often line rock cavities (see pp.56–57).

varieties of quartz

Found in most rock types, quartz is the second most common mineral in Earth's continental crust, after feldspars (see pp.50–51). There are many varieties, and although pure quartz is colourless, the mineral is found in a wide range of colours, which are determined by chemicals present in the rock. Quartz is fairly hard, due to the strength of its chemical bonds, and is the only silicate (oxygen- and silicon-based compound) to be made up entirely of silicon and oxygen. Together with feldspars and mica, quartz is one of the main components of granite. There are two forms of quartz: microcrystalline and macrocrystalline.

Smoky quartz
When quartz specimens, such as the smoky quartz shown here, have space to grow, they form beautiful, hexagonal prisms that terminate in six-sided pyramids. The brown hue of smoky quartz can range from light brown to nearly black, and is produced when the mineral is exposed to radiation underground.

Apple-green colour is due to the presence of nickel

CHRYSOPRASE

Red colour, due to traces of iron oxide, is broken up by criss-crossing white quartz veins

RED JASPER

Brittle, flat surface of green and brown microcrystals

GREEN JASPER

Straight, parallel, alternating white and brown bands

ONYX

Crystal surface has vitreous (glassy) lustre

Striations (parallel grooves) are present on some of the crystal faces

Long, well-formed, six-sided prisms grow in a cluster

One of six faces that top the quartz prisms

SILICATE GROUPS

All silicate minerals (including feldspars) share a basic unit: the silica tetrahedron, in which a silicon atom is surrounded by four oxygen atoms. These tetrahedra, which can be visualized as pyramids or as a ball-and-stick model, occur singly or joined up in arrays. The many silicate minerals are divided into groups according to how the tetrahedra, and other elements, are linked together.

THE SILICA TETRAHEDRON

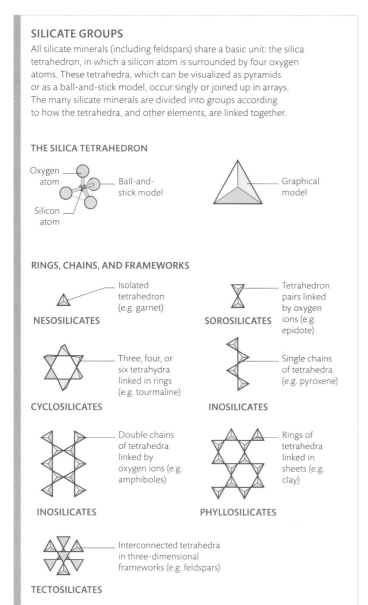

Oxygen atom

Ball-and-stick model

Silicon atom

Graphical model

RINGS, CHAINS, AND FRAMEWORKS

Isolated tetrahedron (e.g. garnet)

NESOSILICATES

Tetrahedron pairs linked by oxygen ions (e.g. epidote)

SOROSILICATES

Three, four, or six tetrahydra linked in rings (e.g. tourmaline)

CYCLOSILICATES

Single chains of tetrahedra (e.g. pyroxene)

INOSILICATES

Double chains of tetrahedra linked by oxygen ions (e.g. amphiboles)

INOSILICATES

Rings of tetrahedra linked in sheets (e.g. clay)

PHYLLOSILICATES

Interconnected tetrahedra in three-dimensional frameworks (e.g. feldspars)

TECTOSILICATES

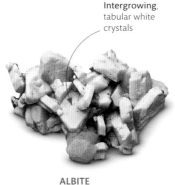

Intergrowing, tabular white crystals

ALBITE

Blue-green amazonite is intergrown with other minerals

AMAZONITE

Outer crust does not show labradorescence as the twinning surface is not exposed

feldspars

The most abundant of all minerals, feldspars make up about two-thirds of Earth's crust. They belong to the silicates, which form the largest and most important class of minerals. Although there are many varieties of feldspar, they can be broadly divided into two groups: potassium-rich feldspars, and sodium- and calcium-rich (collectively known as plagioclase) feldspars. Potassium-rich feldspars are often found in igneous rocks such as granite, as well as in gneiss and sandstone. Plagioclase feldspars are common in igneous rocks such as gabbros, and in meteorites and moon rocks.

Labradorite
A plagioclase feldspar, labradorite is easily identified by its iridescent sheen. The source of this iridescence (or labradorescence) is the mineral's internal structure composed of lamellar twinning – alternating layers of two or more feldspars of different chemical compositions. When light penetrates the mineral, it reflects off these different twinning surfaces, producing flashes of colour.

Tablet-shaped, pale pink crystal

ORTHOCLASE

Prism-shaped, green crystals

ANORTHITE

Greyish, clear-looking crystals with no defined shape

BYTOWNITE

Creamy, salmon-coloured crystals and a massive habit

OLIGOCLASE

Distinct, parallel lines (striation) caused by twinning

Blue and golden iridescent sheen

Specimen is breaking with uneven fracture surfaces

Malachite

A secondary mineral, malachite is formed when copper-rich minerals are altered. This bright green mineral forms when carbonated water (water containing dissolved carbon dioxide) interacts with copper or when limestone interacts with a copper-rich fluid. It often forms round masses of crystals arranged in concentric bands.

In this specimen, a copper sulfide mineral has been overgrown and replaced by a crust of malachite crystals

Individual crystals are short and needle-like

altered minerals

Minerals are chemical compounds that can undergo reactions with other compounds to form new minerals called secondary minerals. When chemical processes change a mineral's chemistry or the properties of its crystals, that mineral is said to be altered. Minerals can be altered when they come into contact with chemicals dissolved in water. This can happen, for example, when existing minerals come into contact with oxygen-rich groundwater percolating down from the surface or with water that has been heated below ground by a body of magma.

Surface has metallic lustre and iridescent tarnish

Chalcopyrite

A copper iron sulfide, chalcopyrite is the main source of copper in the world. When it is exposed to air, its surface tarnishes to iridescent purple, yellow, blue, and green. When chalcopyrite reacts with some water solutions, it can be altered to other minerals such as malachite.

Malachite crystals form tufts

DIAGENESIS

Minerals can be altered into new minerals when sediments are deposited and buried and then consolidated into rock. This process, which takes place near Earth's surface, is known as diagenesis. It can also involve the progressive bonding together of grains of sediment, called cementation. For example, quartz can grow between sand grains, cementing them together.

Pore space

Sand grain

1. SAND

Quartz cement starts to grow between grains

Most grains are cemented

Some pore spaces remain

2. CEMENT GROWTH

Grains glued together by cement

Cement fills all pore spaces

Cemented grains form solid rock

3. SANDSTONE

mineral associations

Minerals frequently occur in groups, or associations, and certain minerals are often found together. For example, malachite is often found in association with pyrite and chalcopyrite, and white gold with quartz. Some minerals form together in a specific environment. Ore minerals, for example, can develop in rocks surrounding cooling magma bodies in Earth's crust (see panel, below). Other mineral associations are related to the chemical composition of the fluids that precipitate them, like the minerals found lining crystalline cavities such as geodes (see pp.56–57). Knowing the environment of formation and the mineral associations helps with the identification of the minerals.

Prehnite with apophyllite
The minerals prehnite and apophyllite are frequently found together, lining cavities inside igneous rocks, such as basalt, or sometimes growing on granite. These two minerals are also found together in veins within rocks (see panel, below). Individual crystals of prehnite are rare. It more usually forms as botryoidal or rounded crystal aggregates.

Minerals may be intergrown, showing they grew around the same time

Pyrite with sphalerite and quartz
This specimen shows colourless quartz with brassy coloured pyrite and dark grey, metallic sphalerite.

Clear crystals of apophyllite

The rock on which the crystals grow is called the matrix

HYDROTHERMAL VEINS

Mineral associations are often found in deposits in Earth's crust called hydrothermal veins. These form by precipitation of minerals from hot, mineral-bearing liquids moving through fissures inside rocks. The liquids can be released by cooling magma bodies called intrusions, or by heated groundwater. The deposits often include economically valuable ore minerals, with tin and tungsten formed close to the intrusion, and copper, zinc, and lead further away.

Weathering turns ore minerals at surface into different minerals

Mineral deposits form in small cracks or permeable zones adjacent to veins

Hot groundwater enriched with minerals

Hot, mineral-rich liquid left from crystallization of granite intrusion

Fluids can reach the surface as geysers or hot springs

Surface water seeps down through the layers

Hydrothermal fluids rise through fractures or fissures and migrate along bedding planes

Apophyllite forms
clusters of blocky,
tabular crystals

Rare, pinkish orange
prehnite crystals have
a glassy lustre

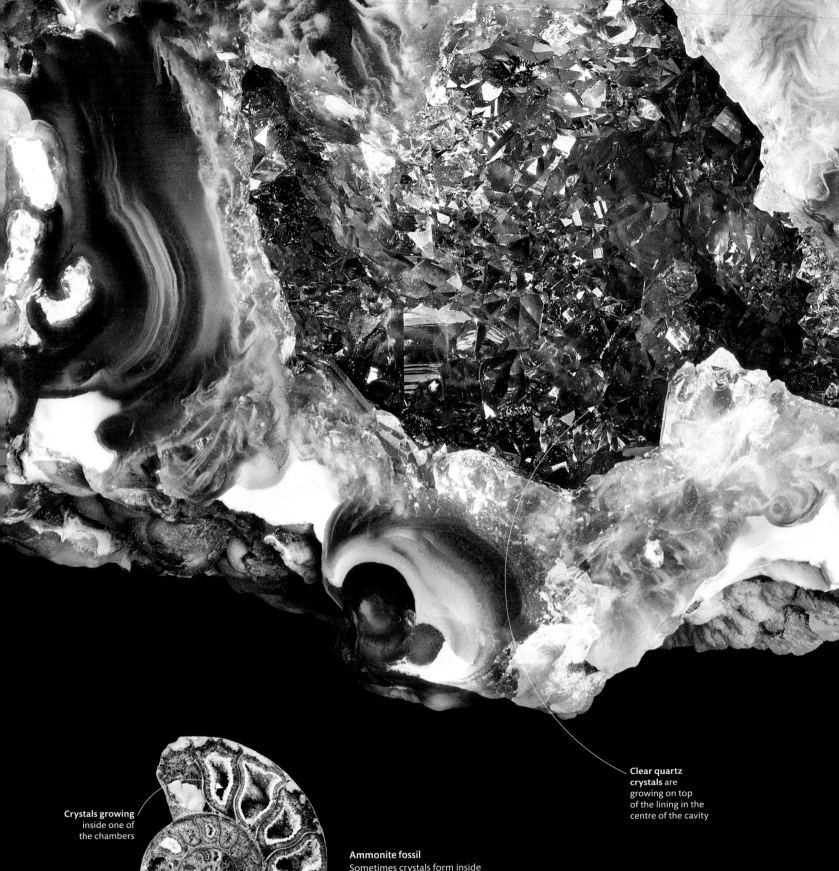

Clear quartz crystals are growing on top of the lining in the centre of the cavity

Crystals growing inside one of the chambers

Ammonite fossil
Sometimes crystals form inside the remains of animals preserved as fossils. For example, fossilized ammonites – extinct relatives of squid and cuttlefish – may contain cavities ("chambers") in which the organism once lived. When mineralizing fluids interact with the chambers, crystals can develop

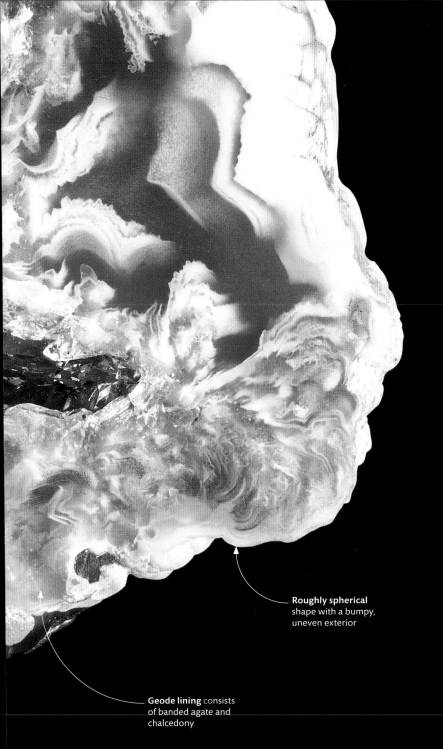

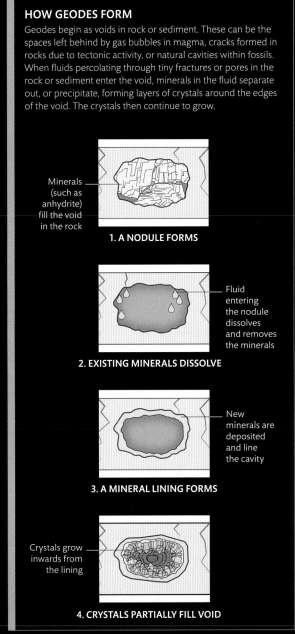

HOW GEODES FORM

Geodes begin as voids in rock or sediment. These can be the spaces left behind by gas bubbles in magma, cracks formed in rocks due to tectonic activity, or natural cavities within fossils. When fluids percolating through tiny fractures or pores in the rock or sediment enter the void, minerals in the fluid separate out, or precipitate, forming layers of crystals around the edges of the void. The crystals then continue to grow.

Minerals (such as anhydrite) fill the void in the rock

1. A NODULE FORMS

Fluid entering the nodule dissolves and removes the minerals

2. EXISTING MINERALS DISSOLVE

New minerals are deposited and line the cavity

3. A MINERAL LINING FORMS

Crystals grow inwards from the lining

4. CRYSTALS PARTIALLY FILL VOID

Roughly spherical shape with a bumpy, uneven exterior

Geode lining consists of banded agate and chalcedony

crystalline cavities

A geode is a hollow, usually rounded rock structure that from the outside resembles a nondescript nodule. However, when split open the geode reveals a cavity lined with minerals, which typically grow into beautiful crystals that project inwards. Unlike crystals growing in confined conditions, crystals inside a geode usually have enough space to develop well-formed faces. As a result, they are highly sought after. Geodes range in size from a few centimetres to several metres across.

Quartz geode
This geode, found in northern Bohemia in the Czech Republic, is filled with quartz varieties, including chalcedony, banded agate, and clear crystalline quartz. Other minerals that grow in geodes include the quartz varieties amethyst and jasper, calcite, dolomite, and celestite. The lining can be made of successive, concentric layers, giving the geode a banded appearance.

HARDNESS

A gemstone's hardness is a key factor when it comes to jewellery, since the cut stones must survive abrasion from repeated use. The Mohs scale (see p.42) is a hardness test that rates minerals in order of their "scratchability" on a scale of 1 (very soft) to 10 (very hard). Ideally, gemstones are at least as hard as quartz (7 on Mohs scale). Diamonds are by far the hardest gemstone.

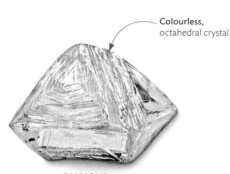

Colourless, octahedral crystal

DIAMOND
10 (Mohs scale)

Cluster of orange-red crystals with a vitreous lustre

GARNET
7–7.5 (Mohs scale)

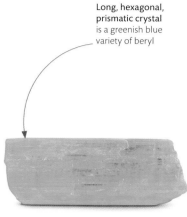

Long, hexagonal, prismatic crystal is a greenish blue variety of beryl

AQUAMARINE
7.5 (Mohs scale)

COLOUR

In diamonds, clarity and an absence of colour make for the most desirable gemstone. However, in most other cases, specific colours make gemstones highly desirable and add to their value. When beryl is colourless, it is of only moderate value, but emeralds (see opposite) – the green variety of beryl – are among the world's most sought-after gemstones.

Purple crystals get their colour from traces of iron

AMETHYST

Green variety of olivine is coloured by iron

PERIDOT

Reddish, tablet-shaped crystal (variety of corundum)

RUBY

Blue, six-sided bipyramidal crystal (variety of corundum)

SAPPHIRE

Orange-brown crystal fragment, with perfect cleavage surface and vitreous lustre

TOPAZ

Different-coloured layers are produced by traces of different chemicals

TOURMALINE

Tough, interlocking, granular texture and a greasy to vitreous lustre

JADEITE JADE
7 (Mohs scale)

Sky-blue nodular tourmaline, with dark veining and no visible crystals

TURQUOISE
5.5–6 (Mohs scale)

Multicoloured blue and purple crystal fragment; although soft for a gemstone, fluorite is prized for its exceptional colours

FLUORITE
4 (Mohs scale)

Emerald
Together with diamond, sapphire, and ruby, emerald is one of the world's most valuable gemstones. Emeralds are found all over the world, but the three main producers are Colombia, Brazil, and Zambia. An alternative name for emerald is green beryl – the silica-rich mineral from which it derives.

Six-sided emerald crystal with typical rich green colour

gemstones

The three key attributes that qualify a mineral to be classed as a gemstone (a jewel or stone used in jewellery) are durability, beauty, and rarity. The beauty of gemstones is enhanced by polishing and cutting the faces of the stones at specific angles to maximize their brilliance and colour. Gem dealers assess gemstones based on their clarity, how they are cut, their colour, and their weight (measured in carats). Just 130 minerals – or fewer than 4 per cent of all minerals – are considered gemstones. In addition, some non-crystalline minerals of organic origin, such as pearl, coral, and amber (see pp.60–61), are classed as gemstones.

HOW A PEARL FORMS

When molluscs sense the presence of foreign bodies, such as sand particles, between their shell and mantle (the organ that secretes the shell), they form a defensive coating around the object. The mantle epithelium – the layer of tissue cells that surrounds molluscs' bodies – starts to coat the object with concentric layers of aragonite and conchiolin, called nacre (or mother-of-pearl). Over time, the layers enclose the object, forming a pearl.

Mantle | Foreign body

Mantle epithelium

Nacre (mother-of-pearl) | Shell

1. COATING SECRETED

Nacre surrounds foreign body

2. PROTECTIVE LAYER GROWS

Foreign body is totally encased

3. PEARL IS FORMED

Minerals from living organisms

Organic minerals are abundant in nature. The soft-bodied parts of bivalves (such as mussels, clams, and oysters) and gastropods (which include snails, whelks, and conches) build their own homes by secreting calcium carbonate – a chalky mineral – to make shells. Coral skeletons are also made of calcium carbonate, secreted by organisms called polyps. The remains of trees that become buried and heated may become organic materials such as coal, anthracite, and jet.

Layers of calcium carbonate secreted by gastropods form a spiralling shell

GASTROPOD SHELL

Dark specks trapped inside the amber could be plant fragments

organic minerals

Some minerals are formed by living organisms through biological processes. Although these are not minerals in the strictest sense, since they are not created inorganically, they can be referred to as organic minerals. The process by which living organisms produce minerals is called biomineralization, and it results in the formation of materials that include shell, pearl, coral, coal, and amber. Sometimes minerals form in the presence of microbes; for example, stromatolites – which are among the oldest known fossils – are made of sediment trapped by living cyanobacteria.

Nacre (mother-of-pearl) lining the shell interior produces an iridescent lustre

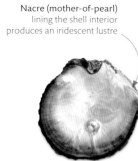

OYSTER SHELL WITH PEARL

Rigid, pink calcium carbonate skeleton secreted by animals called coral polyps, which formed branching colonies

CORAL SKELETON

Shiny black variety of coal with a submetallic lustre

ANTHRACITE

Dark brown to black colour and a woody texture that consists of layers of decomposed trees

JET

Amber

Resin is a thick, sticky substance that is exuded from trees with damaged bark to seal gaps and tears and prevent pathogens and insects from entering. When hardened resin from ancient forests became buried in sedimentary deposits, it formed amber, or fossilized tree resin. Amber can form nodules, rods, and droplets.

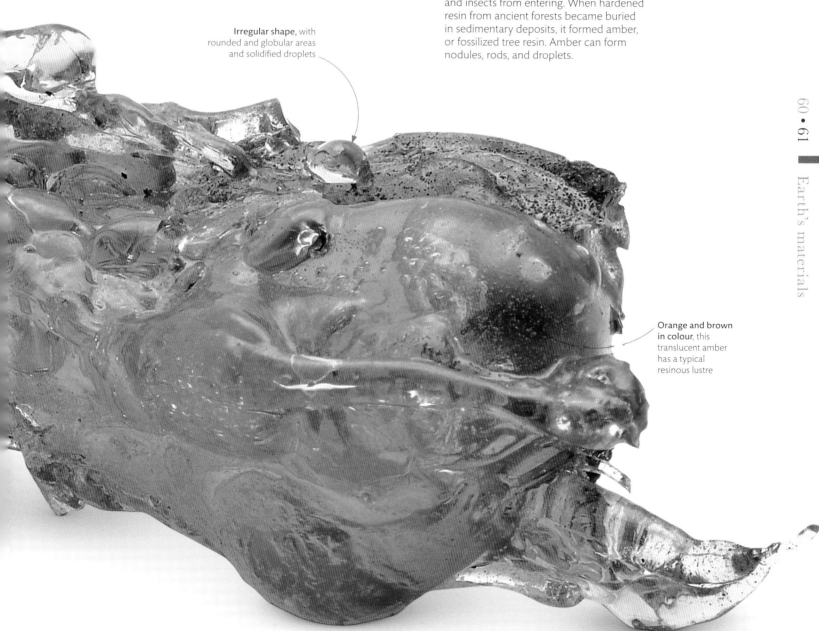

Irregular shape, with rounded and globular areas and solidified droplets

Orange and brown in colour, this translucent amber has a typical resinous lustre

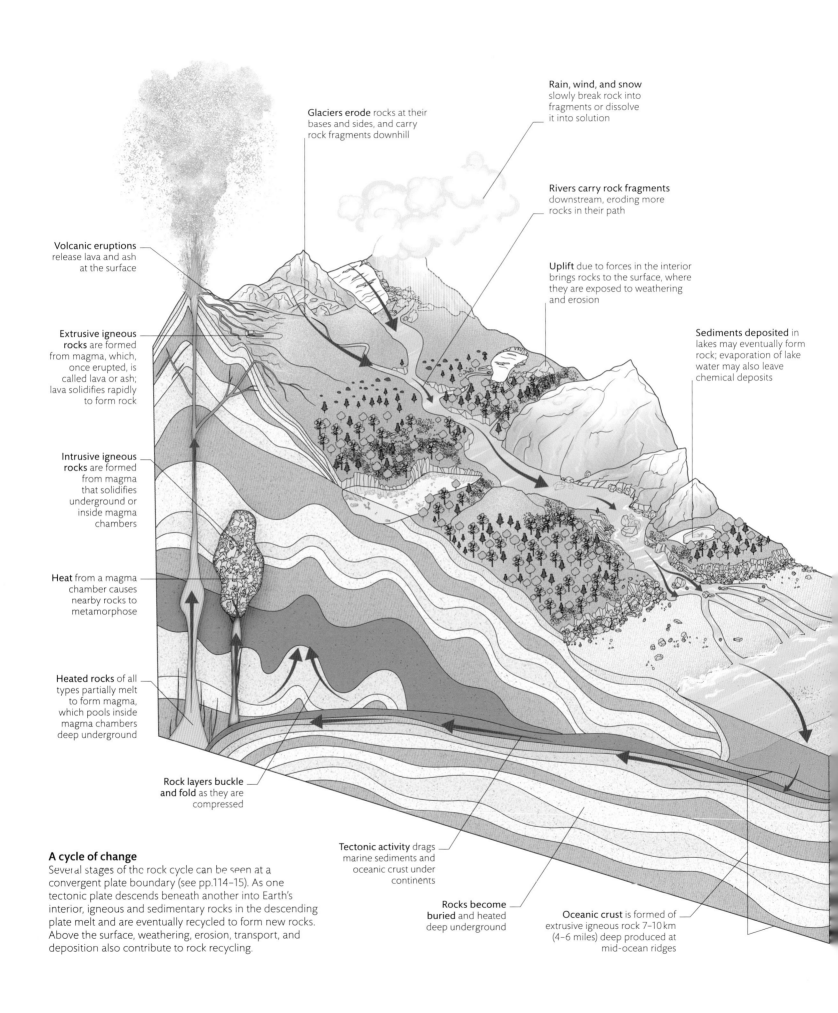

Glaciers erode rocks at their bases and sides, and carry rock fragments downhill

Rain, wind, and snow slowly break rock into fragments or dissolve it into solution

Rivers carry rock fragments downstream, eroding more rocks in their path

Volcanic eruptions release lava and ash at the surface

Uplift due to forces in the interior brings rocks to the surface, where they are exposed to weathering and erosion

Extrusive igneous rocks are formed from magma, which, once erupted, is called lava or ash; lava solidifies rapidly to form rock

Sediments deposited in lakes may eventually form rock; evaporation of lake water may also leave chemical deposits

Intrusive igneous rocks are formed from magma that solidifies underground or inside magma chambers

Heat from a magma chamber causes nearby rocks to metamorphose

Heated rocks of all types partially melt to form magma, which pools inside magma chambers deep underground

Rock layers buckle and fold as they are compressed

A cycle of change
Several stages of the rock cycle can be seen at a convergent plate boundary (see pp.114–15). As one tectonic plate descends beneath another into Earth's interior, igneous and sedimentary rocks in the descending plate melt and are eventually recycled to form new rocks. Above the surface, weathering, erosion, transport, and deposition also contribute to rock recycling.

Tectonic activity drags marine sediments and oceanic crust under continents

Rocks become buried and heated deep underground

Oceanic crust is formed of extrusive igneous rock 7–10 km (4–6 miles) deep produced at mid-ocean ridges

From igneous to sedimentary rock
The Jokulsa A Bru river in the Studlagil Canyon, Iceland, flows between cliffs made of columns of igneous rock that formed from slow-cooling magma. Fragments of the cliffs that are eroded and transported by the river will be deposited downstream or in the sea, where they may eventually become sedimentary rock.

Waves break up rocks at the shoreline and transport the fragments along the coast or out to sea

Rock fragments carried downstream by rivers are deposited along the coast or offshore, with large particles being the first to drop out of suspension

Fine sediments are deposited on the seabed

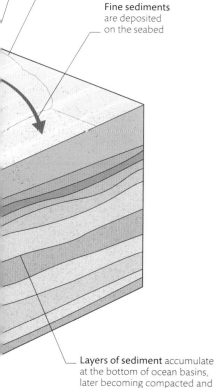

Layers of sediment accumulate at the bottom of ocean basins, later becoming compacted and cemented to form new rock

the rock cycle

Over huge expanses of time, the three basic rock types – igneous, sedimentary, and metamorphic – are constantly changing in a process called the rock cycle. This is governed by the combined effects of heat below the surface, movements of Earth's crust, and surface erosion and deposition. Igneous rocks form when molten rock cools and solidifies below the surface or above ground after being erupted as lava. Sedimentary rocks form when rock fragments, or grains, are loosened or dissolved by the processes of weathering and erosion before being transported, deposited, and then compacted and cemented together. The raw material from which sedimentary rocks form may also be created by animals or plants or by chemical processes. Metamorphic rocks form when igneous or sedimentary rocks are subjected to higher temperatures and pressures that alter their structure or mineral content.

Moss and lichens are growing on the basaltic pillow lava

Rounded shapes of basaltic lava that cooled and formed basalt rock in ancient oceans

Basalt rock, with grooves and ridges

igneous rocks
of the oceans

Earth's thin outer shell is called the crust. There are two distinctly different types: oceanic and continental crust (see pp.66–67). The oceanic crust covers about 70 per cent of Earth's surface and consists of several layers of mainly igneous rock, which combine to form a mass typically 7–10 km (4–6 miles) thick. The oceanic crust is created at mid-oceanic ridges – elevated volcanic structures that form on the ocean floor, at the boundary between two diverging tectonic plates (see pp.110–11). As the plates spread apart, hot magma rises and pools in magma chambers. Some of the magma cools inside the chambers, forming igneous rocks such as gabbro. The rest erupts as lava from fissures on the seafloor and solidifies as basalt.

Pillow basalts
Igneous rocks known as pillow basalts, or pillow lavas, are formed when molten basaltic lava erupts into cold sea water on the ocean floor. The lava cools very quickly and forms a thin skin; with increasing eruptions and pressure from the lava, the skin expands and inflates into rounded, pillow-shaped mounds, typically up to 1 m (3 ft) in diameter. These pillow basalts are in Iceland.

Individual crystals are visible to the naked eye

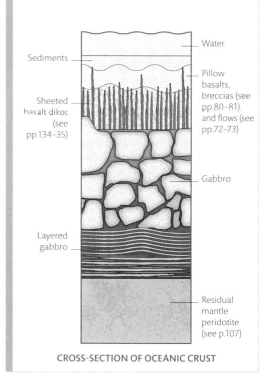

Gabbro
The igneous rock gabbro has the same chemical composition as basalt, but since the magma cools and solidifies more slowly below Earth's surface, there is time for larger crystals to form.

OPHIOLITES
The ocean crust is consistent in structure and thickness, and includes a layered sequence of sediments overlying different rock types. The best place to see oceanic crust is where a section of ancient oceanic crust (called an ophiolite suite or sequence) and the underlying upper mantle have been pushed up onto continental crust, thus exposing it above sea level.

Water

Sediments

Pillow basalts, breccias (see pp.80–81), and flows (see pp.72–73)

Sheeted basalt dikes (see pp.134–35)

Gabbro

Layered gabbro

Residual mantle peridotite (see p.107)

CROSS-SECTION OF OCEANIC CRUST

igneous rocks
of the continents

The continental crust – the part of Earth's crust that makes up the continents and continental shelves – is composed mainly of granite, and is older, thicker, and less dense than the oceanic crust (see pp.64–65). Away from tectonic plate boundaries, the interior portions of continents have remained relatively stable over geological time, since the rocks of the continental crust are relatively buoyant and resist subduction (the process of one tectonic plate sliding under another, which can result in earthquakes and volcanoes). The oldest rocks in the world – about 4 billion years old – are found in the continental crust.

Sharp, jagged peak shaped by weathering and erosion

Ice contributes to weathering as water expands when it freezes, widening existing cracks in rocks

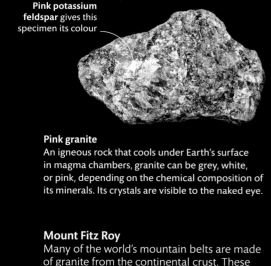

Pink potassium feldspar gives this specimen its colour

Pink granite
An igneous rock that cools under Earth's surface in magma chambers, granite can be grey, white, or pink, depending on the chemical composition of its minerals. Its crystals are visible to the naked eye.

Mount Fitz Roy
Many of the world's mountain belts are made of granite from the continental crust. These rocks formed deep within Earth's mantle. Over time, the granite rocks were uplifted into mountains and were eroded, which is how the jagged peaks of Mount Fitz Roy, in Patagonia – on the border between Chile and Argentina – were formed.

CONTINENTAL CRUST
The continental crust can extend to about 70 km (44 miles) under mountain ranges, although where continents are stretched and thinned, the crust can be as shallow as 20 km (12 miles). It is more varied in composition than oceanic crust, mainly because it is lighter and has not been recycled within Earth's interior to the same extent as denser oceanic crust. As a result, it tends to remain near the surface, where it undergoes repeated cycles of erosion, formation into sedimentary rocks, and metamorphosis.

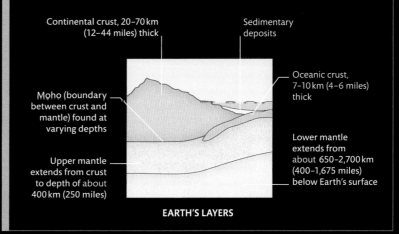

Continental crust, 20–70 km (12–44 miles) thick

Sedimentary deposits

Moho (boundary between crust and mantle) found at varying depths

Oceanic crust, 7–10 km (4–6 miles) thick

Upper mantle extends from crust to depth of about 400 km (250 miles)

Lower mantle extends from about 650–2,700 km (400–1,675 miles) below Earth's surface

EARTH'S LAYERS

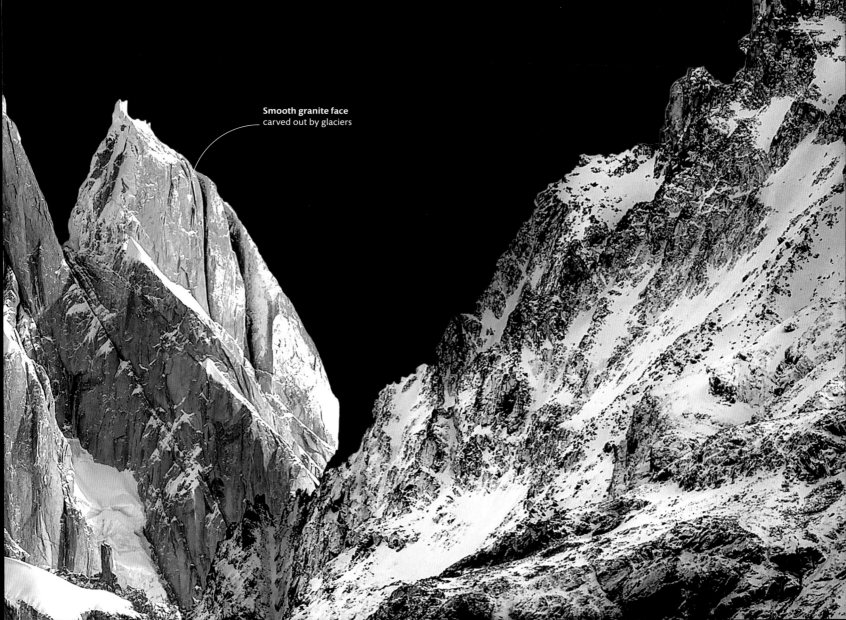

Smooth granite face carved out by glaciers

El Capitan and Half Dome

The steep cliffs of El Capitan tower 1,000 m (3,300 ft) above the floor of Yosemite Valley in California. At the valley's head, the sheer rock face of Half Dome presents an even more spectacular sight. The striking landscape of Yosemite is a result of heat, water, and ice. The granite of El Capitan and Half Dome was intruded as hot magma, deep beneath layers of sedimentary rock, more than 100 million years ago.

Erosion of the sediments had exposed the resistant granite by about 65 million years ago.

Yosemite lies near the crest of the Sierra Nevada Mountains, which were uplifted and tilted from about 25 million years ago, increasing the erosive power of the area's streams, which cut deep canyons into the rock. By about 3 million years ago, the mountain range was high enough for ice fields to form along its

crest when the climate began to cool. Glaciers scoured back the valley's sides, carving out the steep cliff faces.

When the ice receded, a large lake formed on the valley floor, trapped behind the rock debris carried down by the glaciers. The lake eventually filled with sediment, giving the valley its flat floor, which is covered by a mix of forest and meadows. Yosemite Falls, North America's tallest waterfall at 740 m (2,430 ft), is one of several cascading down the valley's steep sides. Yosemite was the first wilderness area to be given legal protection by the US government, in 1864. It became a national park in 1890.

Half Dome

Yosemite valley
Seen here catching the evening sun, El Capitan guards the entrance to Yosemite Valley. On the opposite side, Bridalveil Fall plunges 190 m (620 ft) from a hanging valley. The snow-capped peak of Half Dome can be seen in the distance.

Half Dome rises 2,693 m (8,836 ft) above sea level

Small particles can be teardrop-shaped, spherical, or elongated

Smooth surface with a vitreous lustre (can also be rough or ribbed)

Pele's tears
These small pieces of volcanic glass, up to 20 mm (¾ in) long, are fragments of volcanic rock solidified from low-viscosity lava that cooled rapidly at Earth's surface. They are named after the Hawaiian goddess of fire, Pele.

Lava fountains
Streams of molten lava spew out of fissures in the Kilauea volcano, on Hawaii's Big Island. The lava cools and solidifies into grey-black igneous rock with a jagged, blocky, or rope-like texture.

rocks from lava

Hot molten rock (called magma) is formed in Earth's upper mantle. As it is less dense than the surrounding rocks, it rises up into the crust, where much of it cools and solidifies inside magma chambers in the upper crust. Magma that reaches the surface is erupted from fissures and volcanoes as lava, which then cools and solidifies to form extrusive igneous rock. The explosivity of the eruptions and the rocks that they form depend on the composition and viscosity of the magma. An igneous rock's texture often contains clues to how the rock formed (see panel, below).

IGNEOUS ROCK TEXTURES

When magma cools slowly underground, it has time for large, coarse crystals to form. At Earth's surface, cooling is faster and the resulting crystals are microscopic. Cooling may be so rapid that crystals cannot form and a natural glass is produced. Slow cooling followed by rapid cooling as the magma moves to the surface may result in a porphyritic texture – a mix of large and small crystals. When gas bubbles are present in the lava, they form holes in the rock, giving it a porous appearance. Debris from an explosive eruption forms pyroclastic rocks.

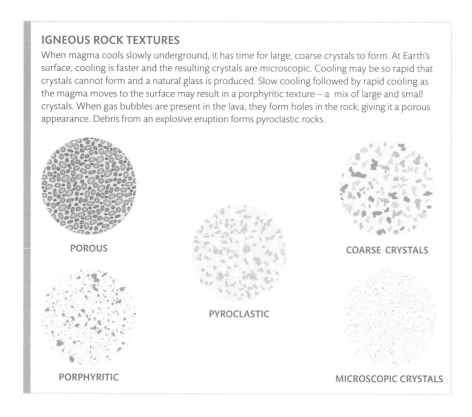

POROUS

PYROCLASTIC

COARSE CRYSTALS

PORPHYRITIC

MICROSCOPIC CRYSTALS

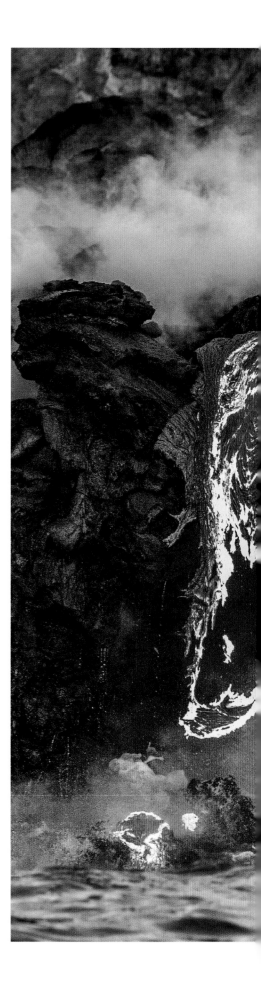

Edge of lava is red hot and is at a higher temperature than the upper surface

Solidified block lava resembles loose rubble

Block lava
As it moves away from a vent, slow-moving lava can cool to form rocks with blocky, angular edges. These block lava deposits were formed on the volcanic island of Nea Kameni, located in the centre of the caldera in Santorini, Greece.

Upper surface changes to a darker, duller colour as it cools

lava

There are many types of lava. A key factor is the amount of silica in the original magma. Magmas that are hot and have low silica content are more likely to rise to the surface and erupt as basaltic lava flows. Because gases escape these magmas easily, they tend to form quiet, or effusive, eruptions. Cooler lavas with a high silica content are thicker. If they reach the surface with a high gas content, they can erupt explosively.

Pahoehoe lava
The surface of a basaltic lava flow can form an elastic skin that is dragged into rope-like folds by the lava underneath. This results in a type of lava called pahoehoe, meaning "ropy" in Polynesian dialect. This close-up image of slow-moving lava in Kilauea, Hawaii, shows a lobe of pahoehoe, which is at more than 1,000 °C (1,800 °F) in places.

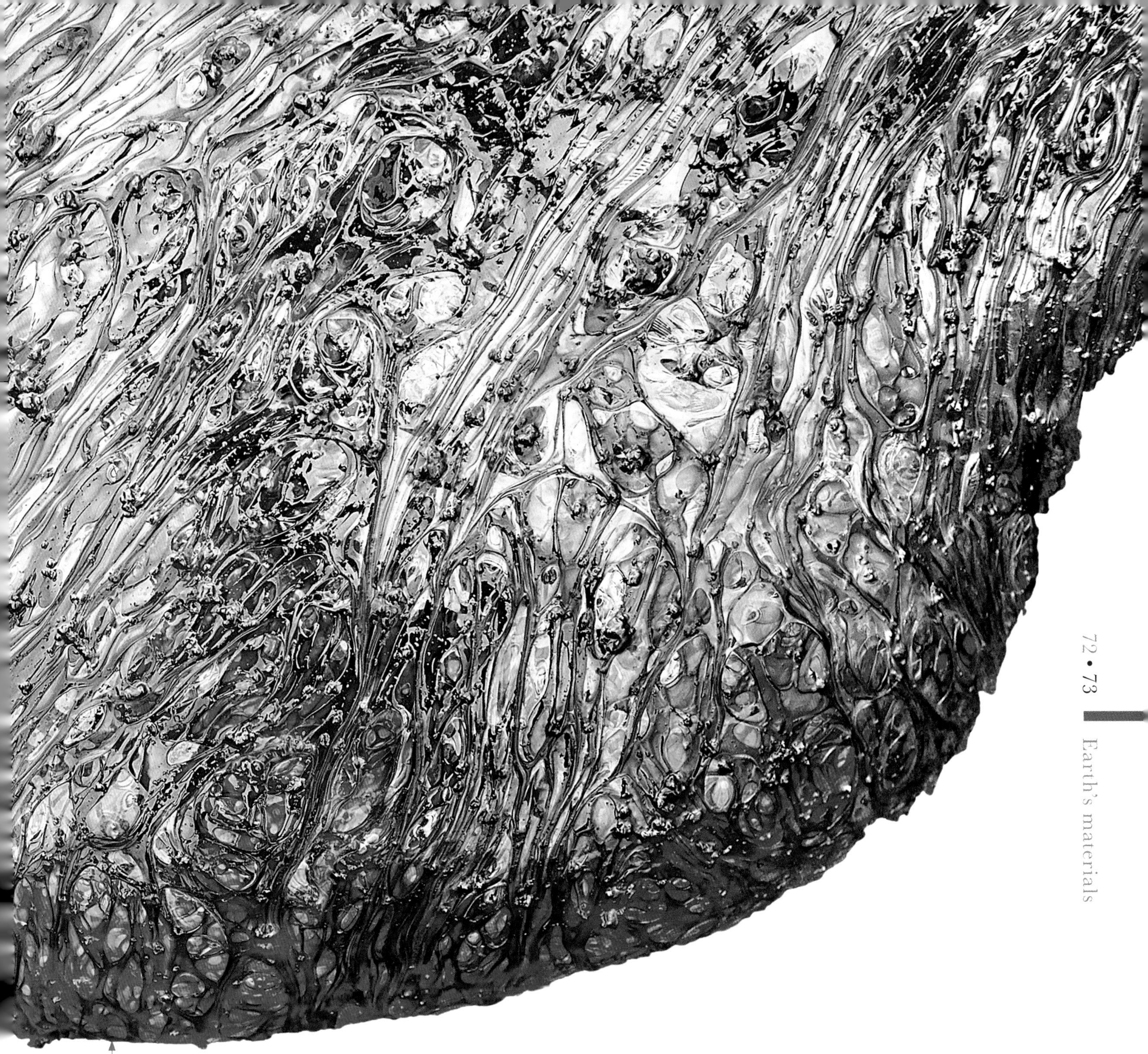

Lava has spread into a
thin, lobe-like sheet at one
edge of a large lava flow

LAVA VISCOSITY

Lava flows can be classified
according to their viscosity,
or resistance to flow, which
increases with silica content.
Lava that is rich in silica can
flow only a short distance
before solidifying, but low-
silica lava can keep flowing for
several kilometres. As viscosity
increases, lava solidifies into
different rocks, from basalt to
andesite, dacite, and rhyolite.

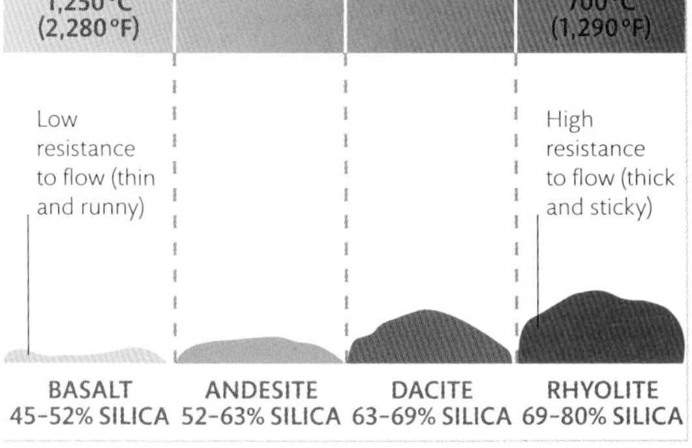

1,250°C (2,280°F)			700°C (1,290°F)
Low resistance to flow (thin and runny)			High resistance to flow (thick and sticky)
BASALT 45–52% SILICA	ANDESITE 52–63% SILICA	DACITE 63–69% SILICA	RHYOLITE 69–80% SILICA

Thin, parallel layers are variable in colour (darker layers were deposited in oxygen-poor environment)

Shale
Sedimentary rocks make up about 5 per cent of the rocks in Earth's crust. Almost 80 per cent of these are fine-grained sedimentary rocks called shale.

Cathedral Gorge
Roughly 1 million years ago, Cathedral Gorge, in Nevada, US, was covered by a freshwater lake. The rock formations that make up the gorge are the layered remnants of the sediments that were deposited in the lake, namely silt, clay, and volcanic ash. Later, erosion and weathering of the rocks by wind and rain created gullies and canyons in the rocks.

fine-grained rocks

Siltstone, mudstone, and shale are fine-grained sedimentary rocks. They consist of tiny particles smaller than sand grains that are deposited in relatively still waters – including lakes, lagoons, swamps, deep ocean basins, and river floodplains – and which become buried and compacted to form parallel layers of rock. These fine-grained rocks vary in texture and hardness. Silt is deposited to form siltstone with a grain size 0.004–0.06 mm in diameter. Clay is deposited to form shale or mudstone made of finer grains (less than 0.004 mm in diameter). Shale splits into thin layers and mudstone breaks into clumps.

Sandy sediment covers floor of gorge

LOESS PLATEAUS
A yellowish brown silt deposit called loess covers over 10 per cent of Earth's land surface. It forms in various environments, including deserts near silt-rich floodplains. The wind carries silt from the floodplain to the desert; fine sediment, like desert dust, being light, is deposited a long way from its source, while coarser, heavier particles are deposited closer. When thick deposits build up, they are called loess plateaus. The largest loess plateaus are in China.

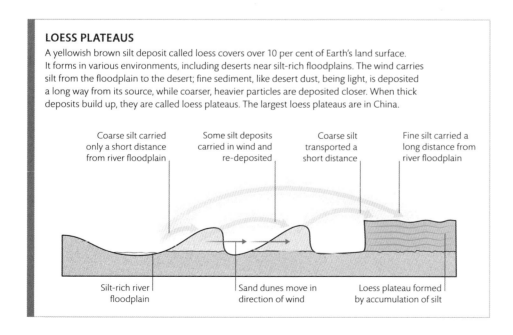

Coarse silt carried only a short distance from river floodplain

Some silt deposits carried in wind and re-deposited

Coarse silt transported a short distance

Fine silt carried a long distance from river floodplain

Silt-rich river floodplain

Sand dunes move in direction of wind

Loess plateau formed by accumulation of silt

Pinnacles and gullies shaped by erosion

Bedding layers are visible among sedimentary rock and ash deposits

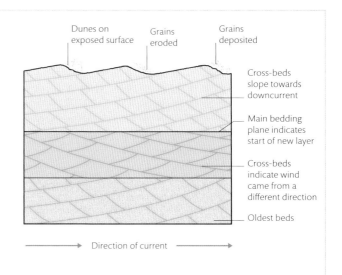

CROSS-BEDDING

Sedimentary rocks form in layers, called bedding. If the sand-sized grains are deposited in still water, the sediment forms layers of rock parallel to the main bedding plane. However, many sandstones display cross-bedding, with grains deposited at an angle to the main bedding plane. Cross-bedding indicates that sediments were transported by a moving current such as the wind, a river, or a stream. The cross-beds in dunes form as sand is carried over the top of a dune and deposited on the other side. As more sand is deposited, sloping layers form.

Dunes on exposed surface

Grains eroded

Grains deposited

Cross-beds slope towards downcurrent

Main bedding plane indicates start of new layer

Cross-beds indicate wind came from a different direction

Oldest beds

Direction of current

sandstone

Sedimentary rocks made up of mainly sand-sized grains (0.125–2 mm/0.004–0.08 in across) are called sandstones. The grains – which consist of minerals, rock fragments, and organic materials such as shell debris and skeletal remains – are transported by wind, streams, rivers, and oceans, and are deposited when the current is no longer strong enough to carry the particles. Over time, the grains become compacted and cemented together to form layers of rock. There are different types of sandstone, depending on the composition of the grains. For example, greywacke, arkose, quartz sandstone, and calcarenite consist mainly of rock fragments, feldspar, quartz, and calcium carbonate respectively. Most sandstone contains high quantities of quartz.

Sea urchin spines, each made of a single crystal of calcite

Navajo sandstone
Antelope Canyon, in Arizona, US, is made from Navajo sandstone, which was formed after sand was deposited by desert winds about 190–170 million years ago. Iron oxides present in the sand give the rock its red colour. Over thousands of years, fast-flowing flood waters eroded the sandstone to form a narrow, steep-sided gorge called a slot canyon. The wave-like shapes were caused by wind-blown sand eroding the rock.

Sand under a microscope
From skeletal remains of organisms to fragments of different rock types, the tiny grains that make up sand display many shapes, sizes, and colours when viewed under a microscope.

cemented fragments

Conglomerates and breccias are sedimentary rocks made of fragments called clasts. In increasing order of size, these clasts range from fine-grained (with the particles larger than sand grains) to pebble- and cobble-sized, with the largest clasts being up to 25 cm (10 in) across. Once the fragments are deposited, they become buried under other rocks and lithify, or transform into solid rock. This process begins with compaction, when the weight of the overlying rocks forces the clasts closer together. When minerals – such as clay, iron oxide, silica, or calcium carbonate – precipitate in the spaces between the clasts, cementation occurs and binds the clasts together. The main difference between breccias and conglomerates is that breccias have clasts with angular edges, while in conglomerates the edges are more rounded. The composition of these rocks gives clues as to the environment and location in which the rock formed.

Conglomerate
This specimen contains jasper and agate, both varieties of microcrystalline quartz (see pp.48–49). The rounded shape and smooth texture of the clasts, typical of conglomerates, indicate that the fragments were deposited far from their point of origin. It is likely they were transported in a fast-flowing river or other high-energy environment capable of carrying the large fragments a long distance and wearing down their sharp edges in the process.

Angular clast set in a finer-grained matrix

Breccia
The relatively angular edges of breccia clasts indicate that the fragments did not travel far from their source. Breccias are often formed at the bases of cliffs or steep hill slopes where weathering debris accumulates.

TYPES OF BRECCIA
There are two types of breccia, depending on the ratio of clasts to matrix (the groundmass of fine-grained particles in which the clasts are cemented). When the clasts touch each other and the matrix fills the remaining voids, the breccia is called clast-supported. When the clasts appear to be "floating" in the matrix and are not in contact with one another, the breccia is called matrix-supported.

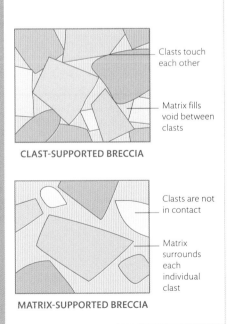

Clasts touch each other

Matrix fills void between clasts

CLAST-SUPPORTED BRECCIA

Clasts are not in contact

Matrix surrounds each individual clast

MATRIX-SUPPORTED BRECCIA

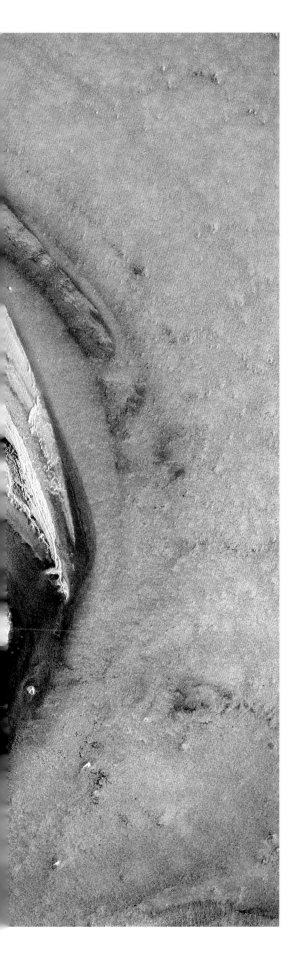

Alternating layers of dark, iron-rich minerals and silica-rich minerals coloured red by iron

Banded-iron formation
About 3.7 to 1.8 billion years ago, oxygen produced by photosynthetic organisms in the ocean reacted with iron dissolved in sea water, creating chemical deposits with a banded appearance (see pp.252–53).

Kati Thanda–Lake Eyre
Spanning about 11,080 square km (4,280 square miles) of desert in South Australia, Kati Thanda–Lake Eyre consists of a thick succession of sediments deposited about 60 million years ago. This aerial view shows the lake bed, which floods periodically, covered in salt crusts left behind by repeated evaporation of a shallow body of mineral-rich water. The bright colours are believed to arise from the presence of bacteria.

HOW SALT CRUSTS FORM

In deserts, lakes containing salt water often evaporate faster than they can be replenished by rainfall (1). When this occurs, precipitation leads to a build-up of salt deposits, including halite and gypsum, in a dry basin called a salt pan (2). Repeated episodes of replenishment and evaporation lead to further layers of salt deposits (3). A hard, dry crust forms on the surface, which may break into polygons.

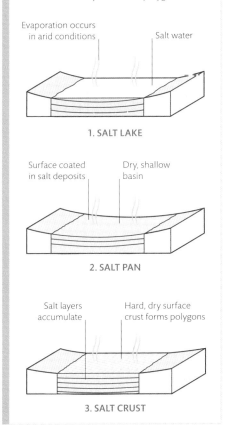

Evaporation occurs in arid conditions

Salt water

1. SALT LAKE

Surface coated in salt deposits

Dry, shallow basin

2. SALT PAN

Salt layers accumulate

Hard, dry surface crust forms polygons

3. SALT CRUST

chemical deposition

Sedimentary rocks formed from fragments of pre-existing rocks and minerals are called clastic sedimentary rocks. Others, called chemical sedimentary rocks, are made of chemicals precipitated out of solutions, often as crystals, and are inorganic or organic in origin. Inorganic rocks formed by chemical deposition include evaporites, dolomite, and inorganic limestone. Organic sedimentary rocks include chert, coal, and organic limestone, mainly formed by marine organisms (see pp.82–83).

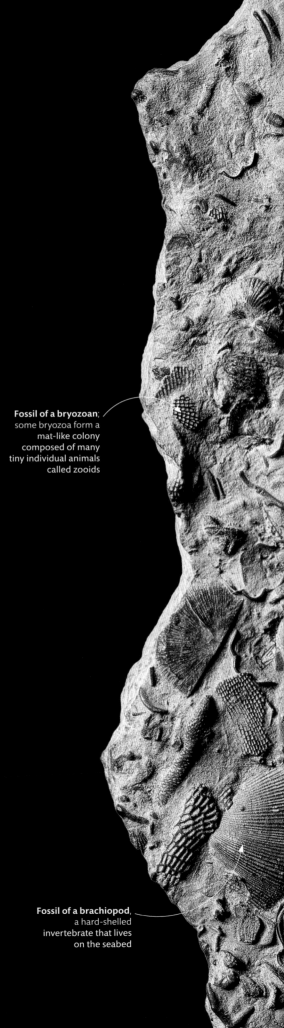

Fossil of a bryozoan; some bryozoa form a mat-like colony composed of many tiny individual animals called zooids

Fossil of a brachiopod, a hard-shelled invertebrate that lives on the seabed

HOW FOSSILIFEROUS LIMESTONE FORMS

Most limestones are formed from the calcium carbonate in the shells and skeletons of marine organisms such as molluscs, sea snails, corals, and crinoids. When the marine organisms (1) die, their hard parts accumulate as calcareous sediment (2). Burial and compaction of these parts in the sediment lead to the formation of limestone (3).

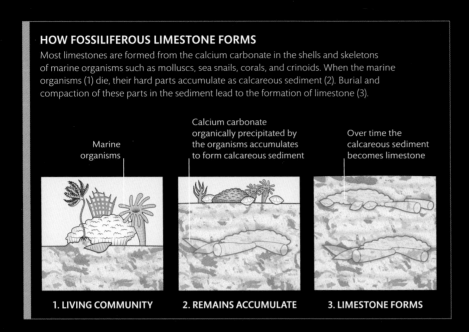

Marine organisms

Calcium carbonate organically precipitated by the organisms accumulates to form calcareous sediment

Over time the calcareous sediment becomes limestone

1. LIVING COMMUNITY   2. REMAINS ACCUMULATE   3. LIMESTONE FORMS

limestone

Accounting for 10–15 per cent of all sedimentary rocks, limestone is composed mainly of calcite (a crystallized form of calcium carbonate). Most limestone is formed from the accumulation of the shells and skeletons of dead marine organisms (see panel, above). However, it is also formed by chemical processes, when calcite is precipitated from sea water or lakes. Caves develop in limestone when slightly acidic water seeps through the rocks and dissolves the carbonate deposits, creating a hole in the rock. When limestone undergoes metamorphism, the calcite recrystallizes to form marble.

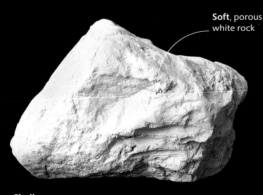

Soft, porous white rock

Chalk
Made from calcite, chalk is a type of limestone formed from the remains of microscopic marine organisms that form a very fine calcareous mud when they die. This mud becomes chalk.

Fossiliferous limestone
Limestones are among the most fossiliferous rocks, with well-preserved fossils providing a useful window on past life. This specimen, which formed about 400 million years ago in a shallow tropical sea, contains abundant fossils of various marine invertebrates, including brachiopods, trilobites, crinoids, and bryozoans.

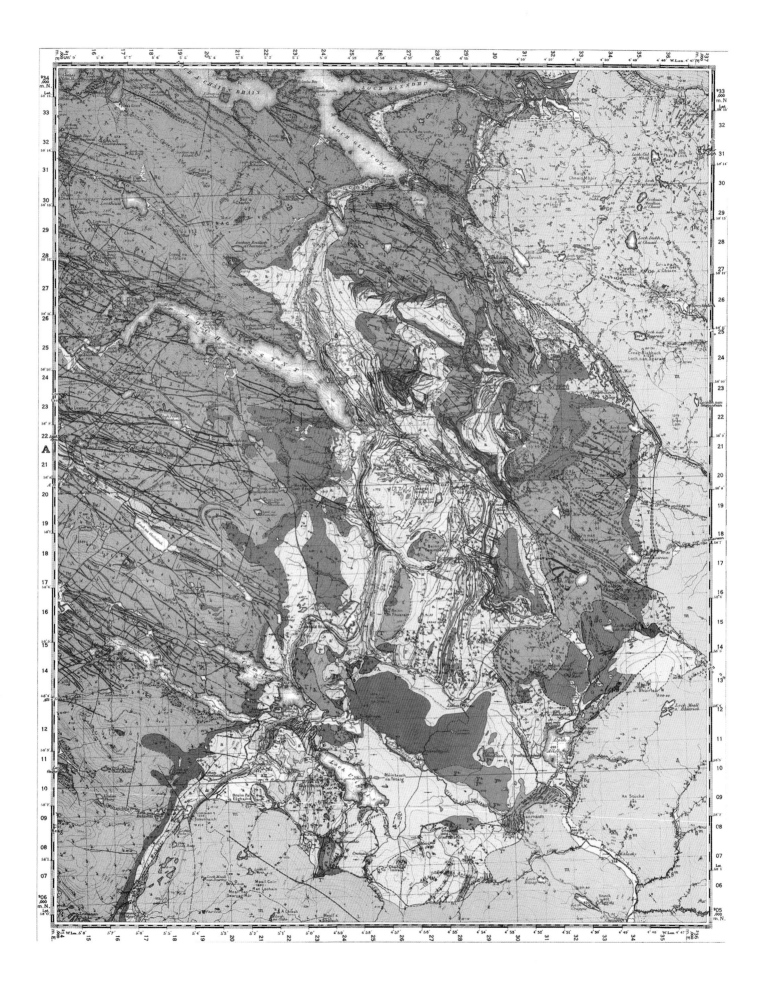

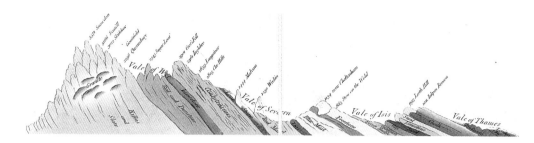

history of Earth science

making maps of rocks

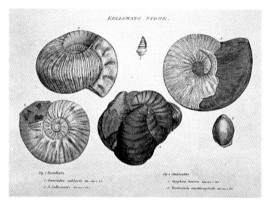

Indicator fossils
This illustration shows some of the characteristic Upper Jurassic fossils found in a pair of rock layers called the Kellaways Formation. William Smith used his understanding of fossils to link outcrops of the same rock layer in different locations.

Geological map of Assynt, Scotland
Compiled at the end of the 19th century, this was one of the first maps to show the complex arrangement of rocks in the heart of an ancient mountain range. As well as using colour to show the rock type, lines represent geological faults and arrows the direction of dip.

For several millennia, maps have been used to record the locations of economically valuable deposits, such as building stones and precious metals – the earliest preserved example is a papyrus map from Egypt, dating from 1150 BCE. But it was not until the 18th century that people started to fill in the gaps between such observations to create systematic and continuous maps of the different rock types beneath their feet.

French geologists Philippe Buache and Jean-Étienne Guettard heralded a new era of map-making when they published a map showing the extent of a layer of chalk across northern France and southern England in 1746. In 1809, Scots-born amateur geologist William Maclure went a step further and produced a geological map of the eastern United States, which classified the surface rocks into four types: primitive, transition, secondary, and alluvial.

Meanwhile, William Smith – a young surveyor in the west of England – had been gaining much more detailed insight into the arrangement of rock layers as he inspected coal mines and canal works. He noticed the same sequence of rocks

as he descended mine shafts in different locations. He also found that these rock layers could be differentiated by the characteristic fossils of ancient flora and fauna that they contained, from which he devised his own principle of "faunal succession".

As Smith supervised canal excavations, he was able to test his ideas across a wide area and to extend his vertical sampling of rock layers into a horizontal representation of their surface outcrops. He started in 1799 with a map of the rock types found in the close surroundings of the city of Bath, where he lived. He spent many more years travelling England, Wales, and southern Scotland, collecting fossils and mapping the rocks in which they were found. In 1815, he produced the first geological map of an entire country, identifying 23 rock layers, each hand-painted in a distinctive colour.

Smith's pioneering work laid the foundations of modern geological mapping. Many of the names he gave to the rock layers he identified are still used by geologists today.

66 Organized Fossils are ... the antiquities of the earth; and very distinctly show its gradual regular formation. 99

WILLIAM SMITH 1817

Serpentinite

The metamorphic rock serpentinite is an example of a rock altered by burial. It is formed when igneous rocks of the oceanic crust are heated to temperatures up to about 200 °C (400 °F) in the presence of water, and their minerals undergo a metamorphic process called serpentinization.

Colours are evenly distributed in this specimen but they may also occurs in bands

Coarse grains can be seen easily with the naked eye

Mylonite
The compact, fine-grained metamorphic rock mylonite (right) is formed by the shearing of rocks along a fault. The parallel banding, or foliation, is typical of rocks produced by dynamic metamorphism.

Bands of tiny, partially recrystallized grains are aligned to the fault

Mottled colouring of yellow and green patches is typical of serpentinite

dynamic and burial metamorphism

Metamorphic rocks are formed in various ways. Dynamic metamorphism occurs when pressure is applied in a particular direction during large-scale movements in Earth's crust, especially along fault planes and along active plate margins. When one part of Earth's crust slides, or shears, past another, the surface along which they slide is called a fault. When faults occur at depth, minerals in the rock partially recrystallize, meaning the rocks become fragmented and re-form to create new rocks with a different texture. Rocks can also be transformed when the temperature and pressure increase due to burial, a process called burial metamorphism.

SHEAR ZONES

At shallow depths in Earth's crust, down to about 10km (6 miles), rocks are relatively cool and brittle. When deformed, they tend to fracture. But in the middle and lower crust, where temperatures are higher, deformation causes rocks to behave like heated plastic and flow rather than break. Over long periods of time, the strain on rocks in these deep locations, called shear zones, causes their minerals to recrystallize and results in the banded textures seen in mylonite (see above).

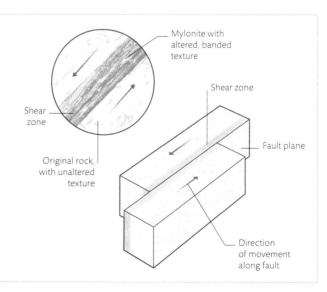

Mylonite with altered, banded texture

Shear zone

Shear zone

Fault plane

Original rock, with unaltered texture

Direction of movement along fault

Gneiss
The metamorphic rock gneiss typically forms
in high-pressure, high-temperature conditions
(over 500°C/950°F). The light and dark banding
is due to chemical reactions that cause the minerals
in the rock to segregate into separate areas with
a different composition. The light areas contain
mainly quartz and feldspar; the dark banding
contains minerals such as pyroxene and amphibole.

Coarse-grained texture
and banded appearance

regional
metamorphism

Rocks that have been altered by pressure or heat, or both, over a large area
have undergone a process called regional metamorphism. This typically
occurs when tectonic plates collide to form mountain belts (see pp.114–15
and pp.120–21). In this situation, rocks in the crust become compressed
(see panel, below), and rocks that were once at Earth's surface can be
dragged tens of kilometres down into the crust, where temperatures are
much higher. The type of metamorphic rock that results depends on the
original rock type, as well as the heat and pressure it is subjected to.

Welsh slate
A fine-grained metamorphic rock, slate is formed
by the effects of heat and pressure on shale.
Since slate can be split into thin layers along its
natural cleavage planes (see panel, below), it is
useful as a roofing and paving material.

FOLDING AND FOLIATION

When rock layers are compressed
due to movements in Earth's crust,
they fold, or bend. Also, minerals
in the rock partially recrystallize
without melting. The new crystals
are aligned perpendicular to the
direction of compressional stress
and parallel to the axial plane of
the fold. This alignment is called
foliation. In the case of slate, the
parallel planes (called slaty cleavage)
are closely spaced and are areas of
relative weakness, which is why the
rock can be easily split into thin
sheets to use as tiles.

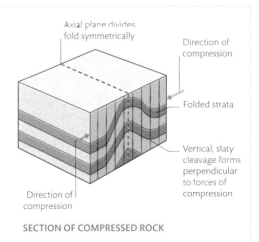

Axial plane divides
fold symmetrically

Direction of
compression

Folded strata

Vertical, slaty
cleavage forms
perpendicular
to forces of
compression

Direction of
compression

SECTION OF COMPRESSED ROCK

contact metamorphism

As intensely hot migrating magma starts to cool and solidify deep below Earth's surface, it forms a body of igneous rock called an igneous intrusion (see pp.134–35). This intrusion releases heat into the surrounding rocks, which undergo a process of chemical change called contact metamorphism. Several types of metamorphic rock are formed in this way, including hornfels – which forms when heat from magma comes into contact with shale, slate, or basalt – and skarn (see right). Since contact metamorphism occurs as a result of temperature increases, rather than pressure, folding and foliation do not occur during the process – unlike in dynamic and regional metamorphism (see pp.86–89).

Rock contains green diopside, pink calcite, and black actinolite minerals

Skarn
When the intense heat from an igneous intrusion reaches carbonate rocks such as impure limestones, the rocks are altered by contact metamorphism. The resulting metamorphic rock is called skarn.

Pre-existing rocks around the dolerite layer have been baked and altered by heat to give a bleached appearance

METAMORPHIC AUREOLES

The area around an igneous intrusion that is affected by contact metamorphism is called a metamorphic aureole. The degree of alteration decreases with distance from the source of the heat, producing a series of distinct concentric zones. In the example shown here, molten magma has solidified to form a granite intrusion. The rocks closest to the magma (hornfels) have undergone greater metamorphosis than those situated further away (spotted rock). The sedimentary rock shale is beyond the aureole and has not been affected by metamorphism.

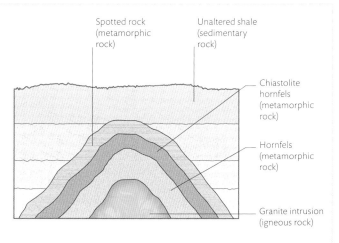

Spotted rock (metamorphic rock)

Unaltered shale (sedimentary rock)

Chiastolite hornfels (metamorphic rock)

Hornfels (metamorphic rock)

Granite intrusion (igneous rock)

Altered by heat

Among the layers of sedimentary rocks that make up the mountains surrounding the Grinnell Glacier in Montana, US, a sill has formed. This occurs when molten magma is thrust between layers of sedimentary rock and solidifies to form a band of igneous rock that follows the sedimentary layers. The intense heat from the sill has altered the rocks immediately above and below through contact metamorphism.

Dark band is a sill made of the igneous rock dolerite – the source of heat that resulted in the alteration of the surrounding rocks

Sedimentary rocks have been deposited in distinct layers over time

Tilted layers
The colourful Rainbow Mountains in north-eastern China are made from sandstone and siltstone that were deposited in the Cretaceous Period, 145–66 million years ago. The layers, which started off as horizontal, were compressed and folded about 50 million years ago during the collision between the Indian and Eurasian continental plates, which resulted in the formation of the Himalayas.

Ravines and gullies in the rocks have been shaped by weathering and erosion from wind and water

Rusty red sandstone is coloured by iron oxide

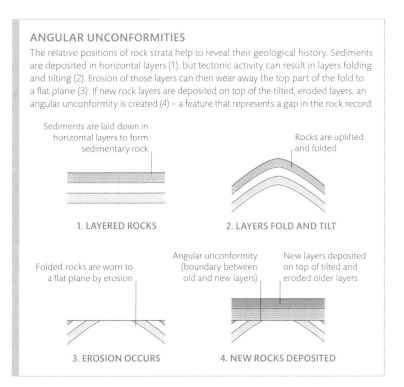

ANGULAR UNCONFORMITIES

The relative positions of rock strata help to reveal their geological history. Sediments are deposited in horizontal layers (1), but tectonic activity can result in layers folding and tilting (2). Erosion of those layers can then wear away the top part of the fold to a flat plane (3). If new rock layers are deposited on top of the tilted, eroded layers, an angular unconformity is created (4) – a feature that represents a gap in the rock record.

Sediments are laid down in horizontal layers to form sedimentary rock

Rocks are uplifted and folded

1. LAYERED ROCKS

2. LAYERS FOLD AND TILT

Folded rocks are worn to a flat plane by erosion

Angular unconformity (boundary between old and new layers)

New layers deposited on top of tilted and eroded older layers

3. EROSION OCCURS

4. NEW ROCKS DEPOSITED

layers of rock

Studying layers of rock allows Earth scientists to understand how and when rocks were formed. Sedimentary rocks are deposited in horizontal layers, with the oldest rocks at the bottom of the sequence and the youngest ones at the top, from which the relative ages of the rocks can be determined. However, the layers can be tilted or overturned by tectonic activity, and erosion can result in gaps in the geological record (see panel, above). By studying fossils embedded in the rock, scientists can assess the ages of the layers, and whether layers that are geographically separated were deposited at similar times.

Each rock layer is younger than the one beneath and older than the one above

Horizontal layers
This cliff (right) on the west coast of Wales, UK, forms part of the group of stratified sedimentary rocks known as the Aberystwyth Grits. The base layer is thought to have been laid down about 488–443 million years ago.

spotlight the Grand Canyon

The Grand Canyon winds across the Colorado Plateau in the southwestern US. Here, the Colorado River has cut 1.6 km (1 mile) through the crust, creating a breathtaking landscape and opening a unique window on 1.8 billion years of Earth's history, revealed in the layers of rocks exposed on its walls. At its widest, the canyon is 29 km (18 miles) across.

The rocks of the Colorado Plateau were deposited from about 575 to 270 million years ago, mostly in shallow seas, beaches, and swamps. The area was then lifted 3,000 m (10,000 ft) above sea level between about 70 and 30 million years ago, probably as a result of a collision between tectonic plates. The canyon itself started to form about 6–5 million years ago when the

Colorado River found an outlet to the sea in the Gulf of California. Falling about 600 m (2,000 ft) over its journey of 446 km (277 miles) through the canyon, the river's course is steep, giving its water tremendous speed and down-cutting power. The water volume and rock load can increase dramatically during times of flood, and would have been much higher at the ends of glacial periods, when upstream glaciers were melting.

The main river has now reached basement rocks that are much harder than the overlying sediments, so the canyon is now growing wider faster than it is becoming deeper, as tributary streams erode its sides.

Aerial view

Steep cliffs are generally made of the most resistant rocks

The canyon at sunrise
The Grand Canyon's cliffs display colourful layers of limestone, sandstone, shale, and lava that overlie a basement of ancient igneous and metamorphic rocks. The uplift of the Colorado Plateau happened without much deformation, leaving the rock strata mainly horizontal and the stratigraphic column easy to read.

Moss is growing on the organic layer on top of the soil

Dark, fertile topsoil, with a high content of organic matter and minerals

Light grey layer is paler than layers above due to leaching of organic matter and iron

Subsoil layer, where leached organic matter and minerals (especially iron oxides, giving the reddish colour) have accumulated; roots also extend down to subsoil

Dark humus layer, which contains live and decomposing plant and animal matter, stores large amounts of carbon

SOIL PROFILES

Most soils are divided into layers, or soil horizons, and their arrangement is called a soil profile – a vertical section through the soil from the surface to bedrock. Not all horizons are present in every soil, but most soils consist of (from top to bottom): a humus layer with plant matter; a dark, fertile topsoil layer; a lighter-coloured, mineral-rich subsoil layer; an infertile, stony layer; and a solid layer called bedrock.

Humus layer, with live and decaying plants, animals, and microorganisms

Dark, fertile topsoil layer contains organic matter and minerals

Lighter-coloured subsoil layer, which is rich in iron, aluminium, and clay minerals

Weathered rock fragments of different sizes

Bedrock, made of solid rock

the soil layer

Comprising a mix of organic matter, minerals, water, air, and living organisms, soils play a vital role in Earth's ecosystems. They absorb and store carbon, recycle dead organisms into nutrients, and provide a key growing medium for plants and a home to a quarter of all known species. Soils also soak up and hold water for plants and other organisms, and control the rate at which excess water drains through. In addition, soils rich in microorganisms break down harmful chemicals, reducing contamination of groundwater.

Parched soil

When drought and dry conditions persist in an area, like this barren plain in Victoria, Australia (right), soils release water and shrink, leading to the formation of cracks. This typically happens in clay-rich soils.

Dry soil has cracked, forming deep fissures

Layers of a podzol

A type of acidic soil called a podzol is distinguished by a light grey layer below the humus or topsoil. It is formed by leaching, when iron and organic matter are washed down into the lower subsoil, usually due to rainfall and rapid drainage. Podzols are found in coniferous forests.

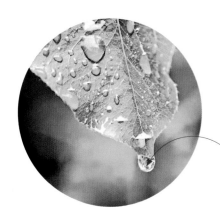

Water droplets
The formation of water droplets, and their roughly spherical shape (left), are due to surface tension, which has a cohesive effect on water molecules interacting at the surface layer (see panel, below).

Water droplet lingering on leaf-tip is held together by surface tension

properties of water

Water is the most abundant chemical in living things. It also has many unusual or unique properties, largely due to its molecules' ability to form hydrogen bonds (a chemical bond that attaches one molecule to another). Few other substances naturally occur at Earth's surface in all three states of matter: as a liquid (water), a solid (ice, see pp.100–01), and a gas (steam or vapour). In addition, water can absorb large amounts of heat without a significant rise in temperature, and releases stored heat when temperatures fall. The relatively constant temperature of water helps to sustain life.

Three states of matter
As the Sun sets over this landscape in Loch Shiel, in the Scottish Highlands, ice is forming a solid surface over the liquid waters of the lake, while water vapour is condensing to form water droplets hanging in the air as mist.

SURFACE TENSION
Water has a high surface tension, which has advantages for some animals and plants – for example, it enables some insects to walk on water and helps transport water from the roots to the leaves of plants. Surface tension is created by the cohesive forces between water's molecules. Within a body of water, the water molecules below the surface are pulled in all directions by surrounding molecules. In contrast, the water molecules at the surface are pulled only inwards or sideways, as there are no water molecules to bond with above the surface. As a result, the molecules at the surface are pulled into the bulk of the liquid, which forces liquid surfaces to contract.

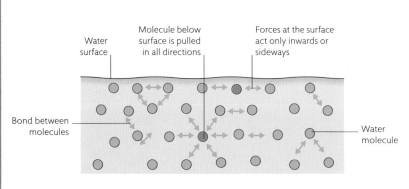

Water surface

Molecule below surface is pulled in all directions

Forces at the surface act only inwards or sideways

Bond between molecules

Water molecule

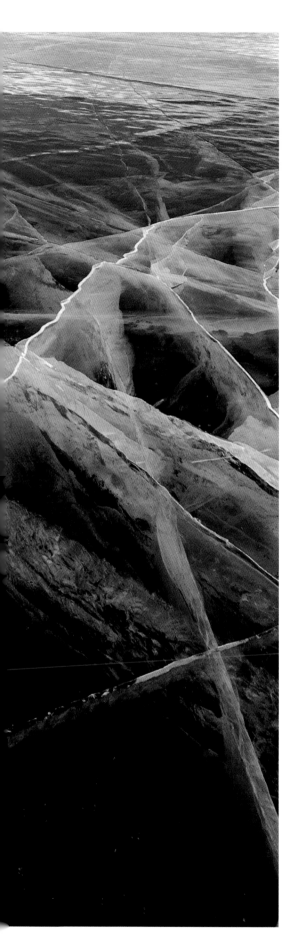

LATENT HEAT

In order for a substance to change state (for example, when a solid such as ice turns into liquid water, or water turns into vapour), heat energy called latent heat is required. The temperature of the substance remains constant during this process. The amount of energy required depends on the state of the substance. Water has high latent heat, since it needs a lot of heat energy to alter its atomic bonds. This means as snow and ice warm, they melt relatively slowly, staying the same temperature until they are liquid.

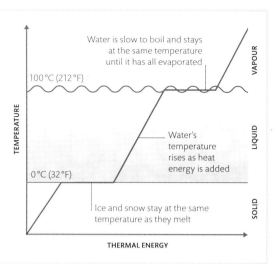

Water is slow to boil and stays at the same temperature until it has all evaporated

100 °C (212 °F)

0 °C (32 °F)

Water's temperature rises as heat energy is added

Ice and snow stay at the same temperature as they melt

TEMPERATURE

THERMAL ENERGY

VAPOUR

LIQUID

SOLID

frozen water

When liquid water cools below its freezing point, 0 °C (32 °F), it starts to turn into solid ice. The molecules in ice are more loosely packed than in water. This means that, unlike most other substances, water expands as it freezes, growing in volume by about 9 per cent. Ice is therefore less dense than water, which is why it floats. Ice is essentially a mineral, being a naturally occurring compound with a defined chemical formula and crystal structure. Snowflakes are ice crystals that have fused together and naturally arranged themselves into intricate hexagonal structures (see pp.230–31).

Lake Baikal

The ice that sits on the surface of the world's deepest lake, Lake Baikal – in Siberia, Russia – is criss-crossed with cracks (left). The natural cracking of ice is caused by daily fluctuations in the temperature of the air, which causes ice to expand during the day and contract at night.

Dendritic, or branching, patterns are typical of frost flowers

Frost flowers

These flower-shaped ice crystals grow on young sea ice (mainly in polar regions) and thin lake ice in calm conditions. They form when water cools rapidly when exposed to much colder air.

Descending ice cloud
Snow and ice hurtle down a steep face of
Baltistan Peak (or K6), Pakistan. Avalanches
such as this one can descend at speeds
of hundreds of kilometres per hour.

avalanches

The sudden movement downslope of a mass of ice or snow,
often incorporating earth and rock, is known as an avalanche.
Avalanches start when a layer of snow or other material becomes
unstable. This can be triggered by snowfall, temperature change,
earthquakes, or disturbance caused by human activities such as
hiking and skiing. There are several types of avalanche. Loose
snow avalanches, also called sluffs, occur when fresh snowfall
moves down steep slopes. Slab avalanches occur when a large
block of solid snow breaks loose and slides down a slope. Powder
snow avalanches comprise components of both loose snow and
slab avalanches. Wet snow avalanches occur when water and
snow move downslope together.

SLAB AVALANCHE FORMATION

These highly destructive avalanches form when heavy, new snowfall accumulates
on a weak, crusty snow layer that is lying on top of older, consolidated snow. The new
snow layer can fracture into blocks or slabs that on a steep slope may slide downhill,
often at great speed. As the slab descends, the front edge disintegrates and forms a
cloud of icy particles, which is sometimes preceded by a powerful blast of air.

Initial fracture

Cracks in the slab

Weak, crusty
snow layer

Cloud of icy
particles and
blast of air

Avalanche
movement

Thick layer of
new snow

Old, compacted
snow

dynamic Earth

Our planet is a complex structure made up of three primary layers. The outermost of these layers is in a state of constant change, driven both by forces in the interior and processes of weathering, erosion, transport, and deposition at the surface. Many of these changes happen too slowly to be perceived in a human lifetime, but occasionally they happen with sudden speed and violence.

Green crystals are omphacite, a sodium-rich variety of the mineral pyroxene

Red crystals are garnets, which is a mineral used as a gemstone (see pp.58–59)

Coarse grains are evenly distributed in this specimen, although they may also form bands

Earth's outermost rocky layer, called the crust, is typically up to 10 km (6 miles) thick under the oceans (oceanic crust) and up to 70 km (44 miles) beneath the continents and continental shelves (continental crust). The thick layer of denser rocks directly below the crust is called the mantle. The solid, brittle top layer of the mantle, which extends down to about 100 km (60 miles) deep, is called the lithospheric mantle. Below this is the asthenosphere, which is made of soft, partially molten rocks and extends to about 350 km (220 miles) beneath the surface. Here, high temperatures and pressure cause the rocks to flow like wax and change shape. Lower down within the mantle, pressure from overlying rocks keeps materials solid. The core, which is composed mainly of iron and nickel, has a liquid outer part and a solid inner part.

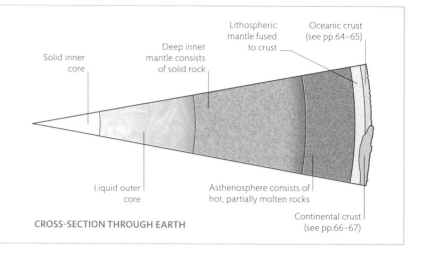

Solid inner core

Deep inner mantle consists of solid rock

Lithospheric mantle fused to crust

Oceanic crust (see pp.64–65)

Liquid outer core

Asthenosphere consists of hot, partially molten rocks

Continental crust (see pp.66–67)

CROSS-SECTION THROUGH EARTH

Large rock fragments from Earth's interior eroded from the mountains towering in the background

Rocks from Earth's interior
The Gros Morne National Park in Newfoundland, Canada (above), provides a rare opportunity to see rocks from Earth's mantle, including peridotite, and oceanic crust. Plate tectonics caused the rocks to thrust through continental crust around 470 million years ago.

Earth's structure

Earth is made up of concentric layers, each of which has different properties and is made of different materials. The three main layers are: a relatively thin outer crust, a large mantle mostly made of the igneous rock peridotite, and a central core; within each layer, there are further subdivisions (see panel, above). Much of what we know about Earth's structure and composition comes from analysing seismic waves – vibrations that are generated by an earthquake, a volcanic explosion, or a landslide and move through different materials in Earth's interior at different speeds. For example, scientists assume that the outer core is liquid because certain types of seismic waves cannot move through liquids. Earth's crust, being closer to the surface, can be sampled directly through drilling.

Garnets are globular, giving eclogite its distinctive spotty appearance

Eclogite
A rare, coarse-grained metamorphic rock, eclogite forms when igneous rocks low in silica, mainly basalt and gabbro, are dragged deep down into Earth's lowermost crust and mantle – sometimes to depths of 150 km (90 miles) – following the convergence of two tectonic plates (see pp.114–15). The intense heat and high pressure cause the igneous rocks to recrystallize and form new rock. Eclogite consists of greenish omphacite and red garnets, but it also contains smaller quantities of other minerals such as quartz and feldspars.

Landsat 9
Orbiting from pole-to-pole once every
99 minutes at a height of 705 km (443 miles),
Landsat 9, launched in 2021, takes more than
700 images per day and revisits each part
of the land surface every 16 days.

history of Earth science

satellites and Earth science

The use of satellites to study Earth and its atmosphere began in 1958 when a research satellite detected Earth's radiation belts. The first successful weather satellite was launched in 1960, and since 1972 detailed images of the land surface have been taken by a series of Earth observation satellites.

Colours represent
differences in height
from a sphere

Gravity "potato"
Low-orbiting satellites are sensitive to small variations in the local gravity field. They have been used to record the shape of the Earth's geoid – the reference surface that the ocean would have without winds and tides. These readings have been used to create visualizations such as the "gravity potato" (above).

Like the early weather satellites, Landsat 1, the first Earth observation satellite, carried a TV camera. It was also equipped with a new multispectral scanner with an infrared channel that enabled it to monitor vegetation growth. Later Landsats added more channels to view more of the spectrum, making it easier to identify different rock types.

During the 1980s and 1990s, other satellites explored different parts of the electromagnetic spectrum: ultraviolet light to track atmospheric ozone; thermal infrared for the temperature of land, ocean, and clouds; and microwave emissions to show water in all its forms – as liquid in the soil, frozen as snow and ice, and as vapour and droplets in the atmosphere. Radar satellites can see through clouds and at night. They are used to measure the height of land and sea, and changes to the thickness and flow of ice sheets.

Satellites allow us to quickly survey remote and inhospitable areas such as the oceans, tropical rainforests, and the polar regions. In the 21st century, geophysical satellites have even opened up a window on Earth's interior, detecting minuscule variations in the planet's gravity and magnetic fields to probe rock density variations in the mantle and the flow of material in the outer core.

Geology from space
This image of the Anti-Atlas Mountains, Morocco, was taken in 2001 by the Advanced Spaceborne Thermal Emission and Reflection Radiometer (ASTER), designed to distinguish different rock types. Layers of limestone, sandstone, and gypsum appear here in yellow, red, and light green, while underlying granites are shown in dark blue and green.

> Man must rise above the Earth ... for only thus will he ... understand the world in which he lives.
>
> ATTRIBUTED TO SOCRATES, 470–399 BCE

Continental crust
Rocks on continental plates are generally old and relatively stable. The oldest continental crust is concentrated in areas known as shields, such as the Australian Shield (see left), in southwest Australia. The rocks in shields are 570 million years old or more.

Oceanic crust
These basalts in the Galápagos Islands formed when the Pacific Ocean's Nazca Plate moved over a plume of hot mantle material. The rising magma welled up to Earth's surface and erupted as lava flow.

Cooled basaltic lava has a twisted, rope-like texture

tectonic plates

Plate tectonics is a widely accepted theory that Earth's solid outer shell, called the lithosphere (made up of the crust and the top layer of the mantle), is divided into large, rigid blocks (plates) that float on the hot, partially molten rocks of the asthenosphere. This causes the plates to move (see panel, below), typically at a rate of 2–20 cm (¾–8 in) per year. The movement of the plates – which may collide, move away from each other, or move past each other – results in the formation and modification of Earth's major surface features, with most activity taking place at the boundaries (see pp.112–13). There are two main types of plates: continental and oceanic, which are about 150 km (90 miles) and 70 km (40 miles) thick respectively.

CAUSES OF PLATE MOTION

Plate motion is thought to be driven mainly by heat rising from the interior of the planet. Convection currents in the mantle lift Earth's lithospheric mantle, and hot, molten magma rises at mid-ocean ridges (see pp.116–17), causing plates to move. The pull of gravity at subduction zones (where a tectonic plate descends into the mantle) and at mid-ocean ridges is also thought to be significant in the movement of plates.

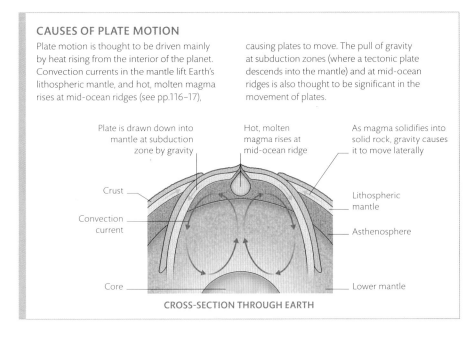

Plate is drawn down into mantle at subduction zone by gravity

Hot, molten magma rises at mid-ocean ridge

As magma solidifies into solid rock, gravity causes it to move laterally

Crust

Convection current

Lithospheric mantle

Asthenosphere

Core

Lower mantle

CROSS-SECTION THROUGH EARTH

TRANSFORM BOUNDARIES

When tectonic plates slide past each other horizontally, in opposite directions, the boundary between them is called a transform boundary or transform fault. This usually occurs on the ocean floor or at the edges of continents. In continental transform boundaries, the continental crust and the lithospheric mantle – the solid, brittle part of the upper mantle fused to the crust – move over the underlying asthenosphere (see below and panel, pp.106–07). The motion between the plates causes the rocks on either side to strain and deform, sometimes resulting in earthquakes.

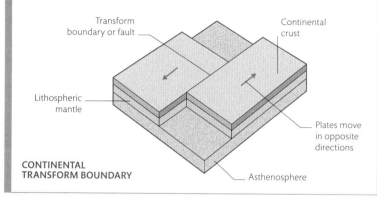

Transform boundary or fault

Continental crust

Lithospheric mantle

Plates move in opposite directions

CONTINENTAL TRANSFORM BOUNDARY

Asthenosphere

plate boundaries

Some of the most intense tectonic activity on Earth takes place at the boundaries of tectonic plates (see pp.110–11). There are three main types of plate boundary: convergent boundaries, where two tectonic plates meet and collide (see pp.114–15); divergent boundaries, where two plates move away from each other (see pp.116–17); and transform boundaries (see panel, above), where the plates slide past one another horizontally. Each type of plate boundary is associated with different geological processes, including mountain-building, volcanic activity, and earthquakes. A single plate can have multiple types of boundary.

Speckled appearance due to pale feldspar mixed with black hornblende and biotite crystals

San Andreas Fault
On the west coast of North America, two adjacent tectonic plates are moving in opposite directions, offsetting hills and landforms. Their transform boundary, which is 1,200 km (750 miles) long, is associated with powerful earthquakes in the region.

Diorite
The igneous rock diorite is associated with convergent plate boundaries. It is an example of an intrusive rock, with visible crystals that solidified slowly in a magma chamber.

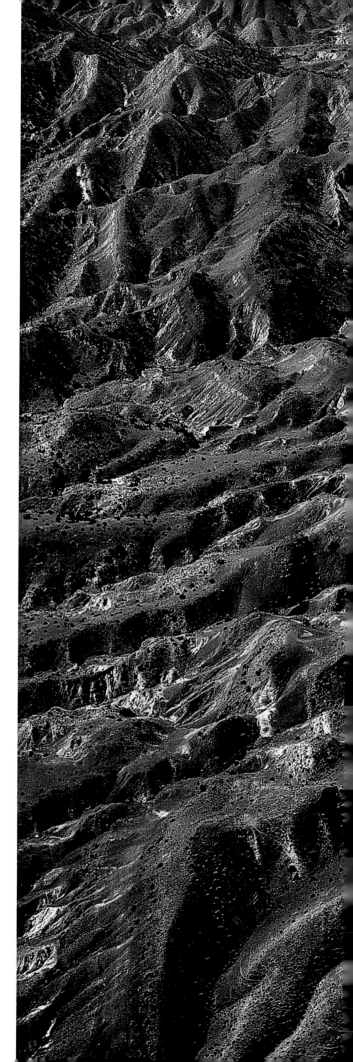

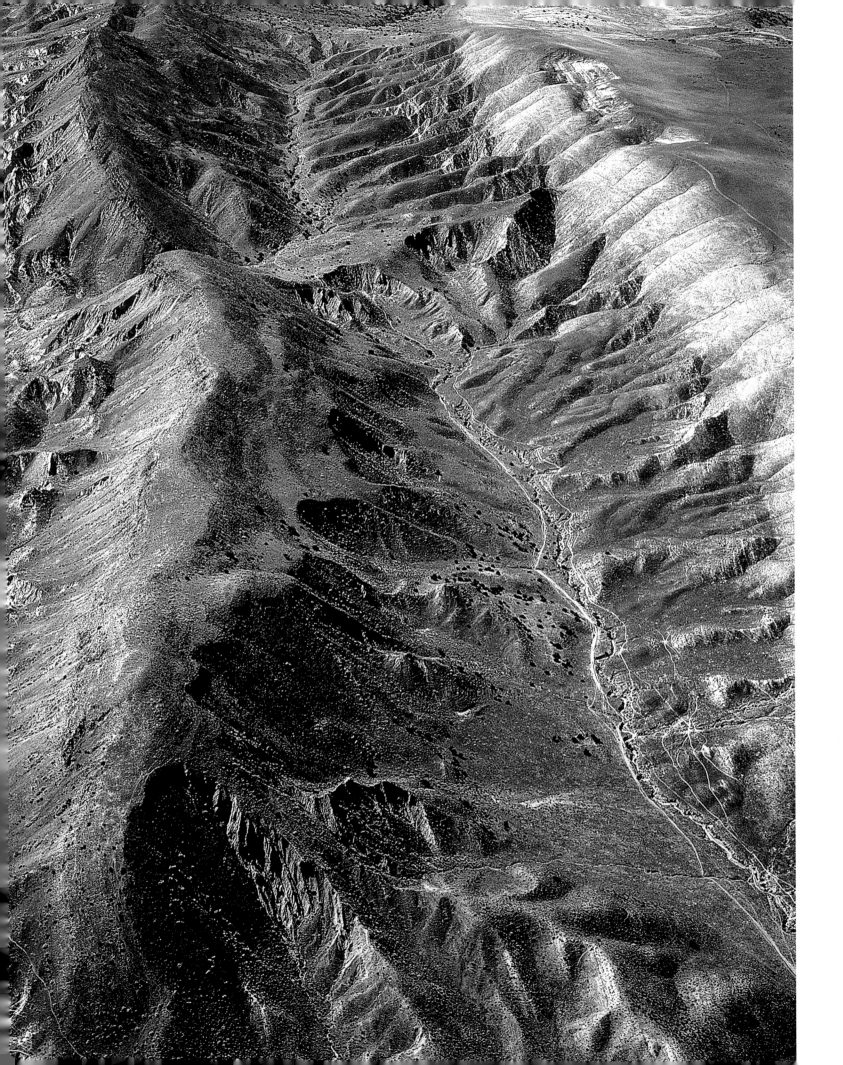

Continent-continent collision
This aerial view shows part of the Ladakh
Range in the western Himalayas. Like the
rest of the range, they are the result of
a collision between sections of continental
crust in the Indian and Eurasian plates.

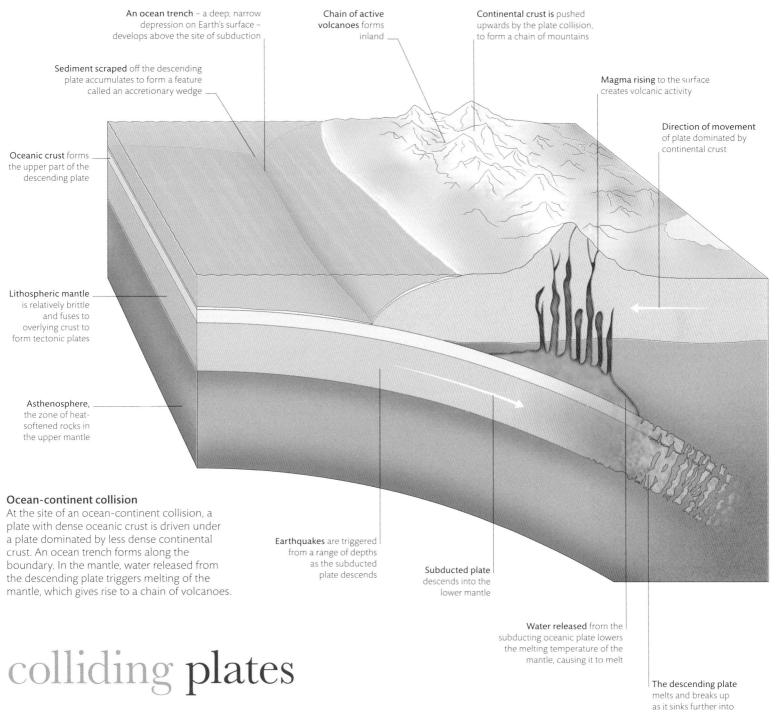

An ocean trench – a deep, narrow depression on Earth's surface – develops above the site of subduction

Chain of active volcanoes forms inland

Continental crust is pushed upwards by the plate collision, to form a chain of mountains

Sediment scraped off the descending plate accumulates to form a feature called an accretionary wedge

Magma rising to the surface creates volcanic activity

Oceanic crust forms the upper part of the descending plate

Direction of movement of plate dominated by continental crust

Lithospheric mantle is relatively brittle and fuses to overlying crust to form tectonic plates

Asthenosphere, the zone of heat-softened rocks in the upper mantle

Ocean-continent collision
At the site of an ocean-continent collision, a plate with dense oceanic crust is driven under a plate dominated by less dense continental crust. An ocean trench forms along the boundary. In the mantle, water released from the descending plate triggers melting of the mantle, which gives rise to a chain of volcanoes.

Earthquakes are triggered from a range of depths as the subducted plate descends

Subducted plate descends into the lower mantle

Water released from the subducting oceanic plate lowers the melting temperature of the mantle, causing it to melt

The descending plate melts and breaks up as it sinks further into the hot mantle

colliding plates

Earth's tectonic plates are constantly moving. When two plates move towards one another and collide at a plate boundary, that boundary is termed convergent. There are three main types of convergent boundary: ocean-continent, where oceanic crust collides with and is forced down (subducted) under continental crust; ocean-ocean, where the denser of two sections of oceanic crust is subducted under the other; and continent-continent, where two continents collide, merging into a single landmass. Because continental crust is of relatively low density, it does not become subducted, but crumples to form mountain ranges. The boundary between two plates in which one is subducted is known as a subduction zone. Where continents collide, the boundary is called a suture zone.

diverging plates

When two tectonic plates stretch, thin, and break apart, the zone where they meet is called a divergent boundary. Where this occurs beneath oceanic crust, magma rising from the mantle forms a deep fissure. New magma then flows into the fissure, spreads outwards, and solidifies to form new crust. If the process (seafloor spreading) is repeated, it can lead to the formation of a submarine mountain range called a ridge (see panel, opposite). Where a divergent boundary occurs under thicker continental plate, the plate fractures to form a rift (trough-shaped valley) with faults either side. Water then flows into the valley from rivers and streams, forming a lake; where the rift drops below sea level, sea water may flow in. If the rift deepens and widens, a new ocean basin can form.

Small lake created when water from rivers and streams entered the valley

Great Rift Valley
In East Africa (including Tanzania, above), valleys and gorges have been created by the thinning of continental crust close to a divergent boundary under the Red Sea and Gulf of Aden.

MID-OCEAN RIDGES

Spreading ridges have different topographies depending on the speed at which the tectonic plates move. In fast-spreading ridges, such as the East Pacific Rise, the ocean floor moves away from the ridge before cooling and subsiding, or sinking, forming broad, gentle landscapes. In slow-spreading ridges, such as the Mid-Atlantic Ridge, the ocean floor does not move far before cooling and subsiding, forming narrow, steep landscapes with faults.

Spreading ridge creates gentle topography

Direction of plate movement

Crust

Lithospheric mantle

Rise of mantle below ridge axis

Asthenosphere

FAST-SPREADING RIDGE

Blocks of crust sink along faults

Steep-sided valley formed

SLOW-SPREADING RIDGE

Oceanic divide

The Mid-Atlantic Ridge extends about 16,000 km (10,000 miles) along the centre of the Atlantic Ocean. The Silfra Fissure in Iceland (shown here), which straddles the Mid-Atlantic Ridge, is one of the few places on Earth where people can dive and swim between two tectonic plates – the Eurasian and North American plates.

Rocks on the limb of the fold, originally part of a syncline, have been compressed and overturned

Smaller folds, called parasitic folds, have formed on the limb of the overturned fold

Mainly limestone, these rocks were deposited on the bottom of an ocean

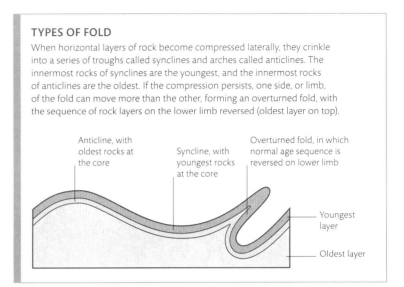

TYPES OF FOLD

When horizontal layers of rock become compressed laterally, they crinkle into a series of troughs called synclines and arches called anticlines. The innermost rocks of synclines are the youngest, and the innermost rocks of anticlines are the oldest. If the compression persists, one side, or limb, of the fold can move more than the other, forming an overturned fold, with the sequence of rock layers on the lower limb reversed (oldest layer on top).

Anticline, with oldest rocks at the core

Syncline, with youngest rocks at the core

Overturned fold, in which normal age sequence is reversed on lower limb

Youngest layer

Oldest layer

Earth's folds

When hot rocks deep within Earth's crust become compressed by the movement of tectonic plates (see pp. 110–11), rather than fracturing or breaking, they can become folded or bent. Folds often form at the boundaries of colliding plates (see pp. 114–15) and can occur on a variety of scales, from a small bend affecting a hand-sized rock specimen to the folds that make up entire mountains, such as in the Swiss Alps and Canadian Rockies. By studying how rocks are folded, geologists can better understand the geological history of an area.

Mountain-sized fold
These layers on the flank of the Dent de Morcles, Switzerland (left), were deposited in the Mesozoic Era (245 to 65 million years ago). They were folded during a collision between Europe and Africa about 65 million years ago.

Fold is symmetrical either side of hinge zone

V-shaped folds
These limestone and chert layers in Crete, Greece, have been folded into V-shapes called chevron folds. These folds have straight limbs that bend abruptly on one point, called the hinge zone.

Snowy spires
The Dolomite mountains of northeastern Italy are still gradually growing, buckling upwards as the Eurasian and African continental plates collide. They are composed of dolomite, a hard carbonate rock, and their dramatic jagged peaks are the result of erosion by rain and ice.

mountain-building

Some mountains are created by the accumulation of basalt that forms from lava erupted from volcanoes on land or on the ocean floor, while others are formed when rivers erode steep gorges into pre-existing plateaus. But one of the most common settings for mountain formation is at the boundary of two continental plates. When two continental plates collide, the crust buckles, thickens, and folds to create high mountains. About 60–40 million years ago a collision between the continents of India and Asia gave rise to the Himalayas (see pp.124–25). Other mountains ranges — for example, the Andes — formed at the boundary between an oceanic plate and a continental plate.

MOUNTAIN ROOTS

Earth's crust is thicker below high mountains than below the surfaces of low elevations. When mountains are built, the crust thickens both above and below the surface. Much like icebergs, mountains have deep roots underneath, which mirror and exaggerate the surface topography. The roots, which are made of a lighter, more buoyant, crustal material, float on the denser mantle underneath. Earth's crust is in a state of buoyant equilibrium called isostasy. According to this principle, the depth to which a floating object sinks depends upon its thickness and density. Over time, the upper surface of a mountain is lowered by the action of water, wind, and ice, and the buoyant root rebounds to balance the loss of mass on the surface.

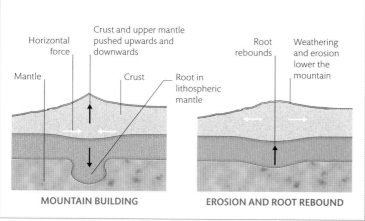

MOUNTAIN BUILDING   EROSION AND ROOT REBOUND

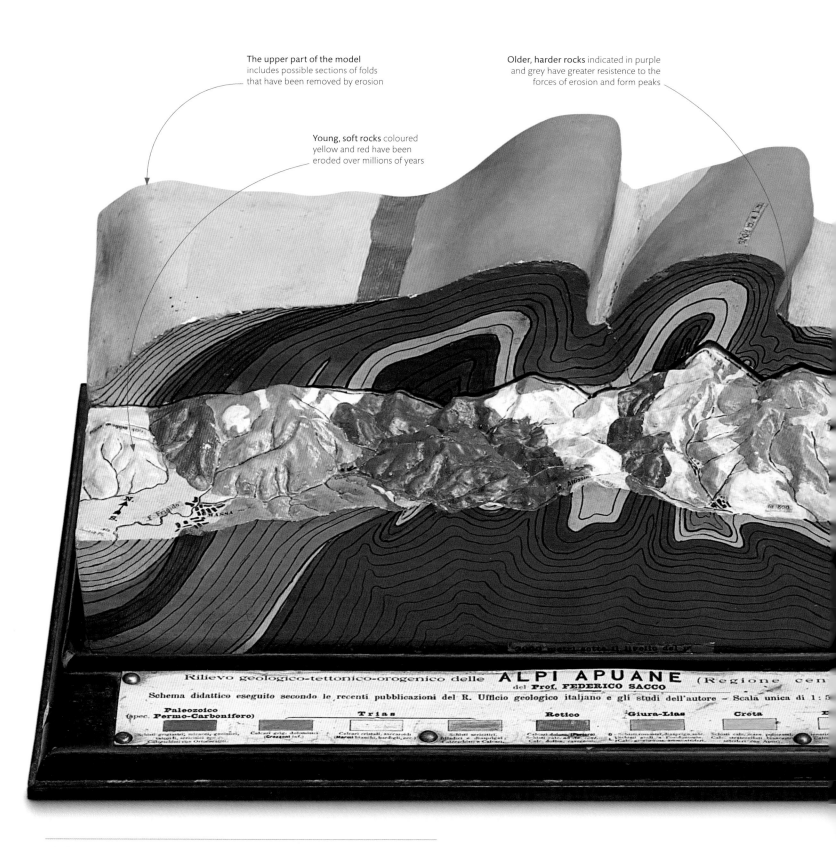

The upper part of the model includes possible sections of folds that have been removed by erosion

Older, harder rocks indicated in purple and grey have greater resistence to the forces of erosion and form peaks

Young, soft rocks coloured yellow and red have been eroded over millions of years

Rilievo geologico-tettonico-orogenico delle **ALPI APUANE** (Regione cen
del Prof. FEDERICO SACCO
Schema didattico eseguito secondo le recenti pubblicazioni del R. Ufficio geologico italiano e gli studi dell'autore - Scala unica di 1: 5

Paleozoico (spec. Permo-Carbonifero)		Trias				Retico	Giura-Lias	Créta	

66 … the work of Suess marks the end of the first day, when there was light. 99

MARCEL-ALEXANDRE BERTRAND, FRENCH GEOLOGIST, 1897

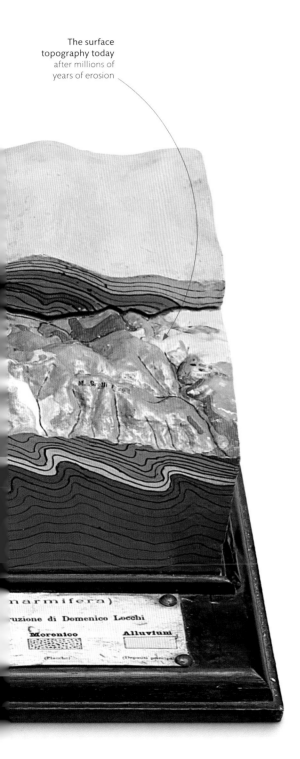

The surface topography today after millions of years of erosion

Relief model of the Apuan Alps
Many early geologists created cross-sectional relief models to explain their ideas. This example was made by the Italian geologist and palaeontologist Federico Sacco (1864–1948).

Albert Heim
Swiss geologist Albert Heim (1849–1937), pictured in later life with local mountain guides in the Maderaner Valley, Switzerland, made pioneering studies of the Alps from the 1880s until the 1900s.

history of Earth science

understanding mountain-building

Much of our understanding of how mountains form is based on the study of the Alps. From the late 1700s, geologists from Switzerland and France began to explore and survey the Alps to produce maps and later three-dimensional models that helped them to understand the structure of the mountains and the processes that gave rise to them.

Swiss scientist and Alpine explorer Horace Bénédict de Saussure (1740–99) made detailed studies of the rocks of the Alps. In the 19th century, further systematic surveys of the rock types of the region were carried out, and in 1853 Arnold Escher von der Linth, a Swiss geologist, completed the first geological map of the whole of Switzerland. But these new maps often raised more questions than they answered. Across the Alps, older rocks were often found overlying younger ones, reversing the expected order of strata.

One of the first to recognize that horizontal movement could be greater than vertical movement in mountain building, Austrian geologist Eduard Suess (1831–1914) concluded that the much-studied Glarus Alps were

the result of a 35-km (21-mile) northward thrust fault that left older rocks on top of younger rocks. He mistakenly believed that the huge horizontal forces were due to the gradual contraction of Earth's crust.

French geologist Marcel-Alexandre Bertrand (1847–1907) took these ideas further, identifying immense thrust sheets across the Alps. He found similar features elsewhere, leading him to identify distinct mountain-building episodes – the Caledonian, Hercynian, and Alpine.

As the Alps became better understood, they became the testing ground for new geological theories. During the 20th century, continental drift and plate tectonics became accepted by geologists partly because they offered better explanations for features that had been studied in the Alps.

the Himalayas

There is no greater expression of the power of tectonic forces than the world's highest mountain chain, the Himalayas, which stretch 2,900 km (1,800 miles) across Central Asia. There are at least ten peaks in the Himalayas over 8,000 km (26,000 ft). No other range in the world has a single summit above this elevation. Here, the fossils of ancient sea creatures can be found on the roof of the world. The summit of Mount Everest consists of marine limestone. These sea floor sediments have not only been lifted 8,848 m (29,032 ft) above sea level but also thrust 2,000 km (1,250 miles) north into the heart of Asia.

About 200 million years ago, the ancient supercontinent of Pangea began to break up. The Indian subcontinent moved rapidly north, colliding with Asia about 60–40 million years ago. Coastal sediments from the shores of both continents were crumpled and piled high to form the Himalayas as the Indian Plate drove under the Eurasian Plate, compressing and

The Karakoram Range

K2 is the highest peak in the Karakoram and the world's second highest mountain

thickening the crust to the north and lifting up the Tibetan Plateau. The plateau is bounded to the west by an extension of the Himalayas, the Karakoram Mountains. The formation of the Himalayas has transformed the region's climate, leading to the establishment of the annual Southeast Asian monsoon. Several major rivers, including the Ganges and the Brahmaputra, also have their sources in the range.

The Himalayas are a young mountain range that is still rising. GPS measurements show that the peak of Mount Everest is rising 1 cm (0.4 in) every ten years, or about 30 cm (1 ft) every 300 years.

The Annapurna Massif
Four peaks carry the name Annapurna. They are part of a massif that stretches 55 km (34 miles) through the central Himalayas. Annapurna I is the highest, at 8,091 m (26,545 ft) above sea level. Its steep sides, unpredictable weather, and the risk of avalanche make it one of the world's deadliest mountains for climbers.

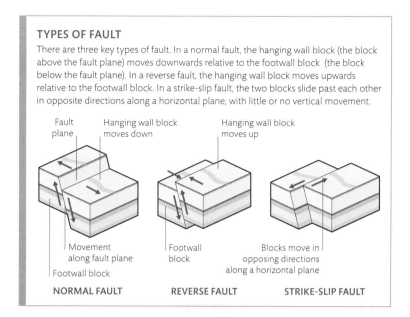
lines of fracture

When blocks of rocks in the brittle outer layers of Earth's crust and upper mantle fracture and move relative to one another, the fracture is called a fault. When the movement is sudden, it generates shockwaves – vibrations that travel outwards through the surrounding rocks – causing earthquakes on Earth's surface. The relative motion between adjacent blocks of rock is called fault movement, and it ranges from a few millimetres to thousands of kilometres. Movement, or slip, takes place along a roughly flat surface called a fault plane (see panel, above). The rock masses on either side of a fault plane are called fault blocks.

This block has moved down relative to the other

Thrust faults

These red and green sandstones and creamy limestones in the Tien Shan Mountains in China (right) have fractured along thrust faults, as older layers of rock have been pushed up on top of younger layers. Thrust faults are a type of reverse fault with movement at relatively low angles of 45 degrees or less.

Fault lines in sandstone

These colourful sandstone beds, which were deposited in parallel layers separated by bedding planes, have been displaced by a normal fault.

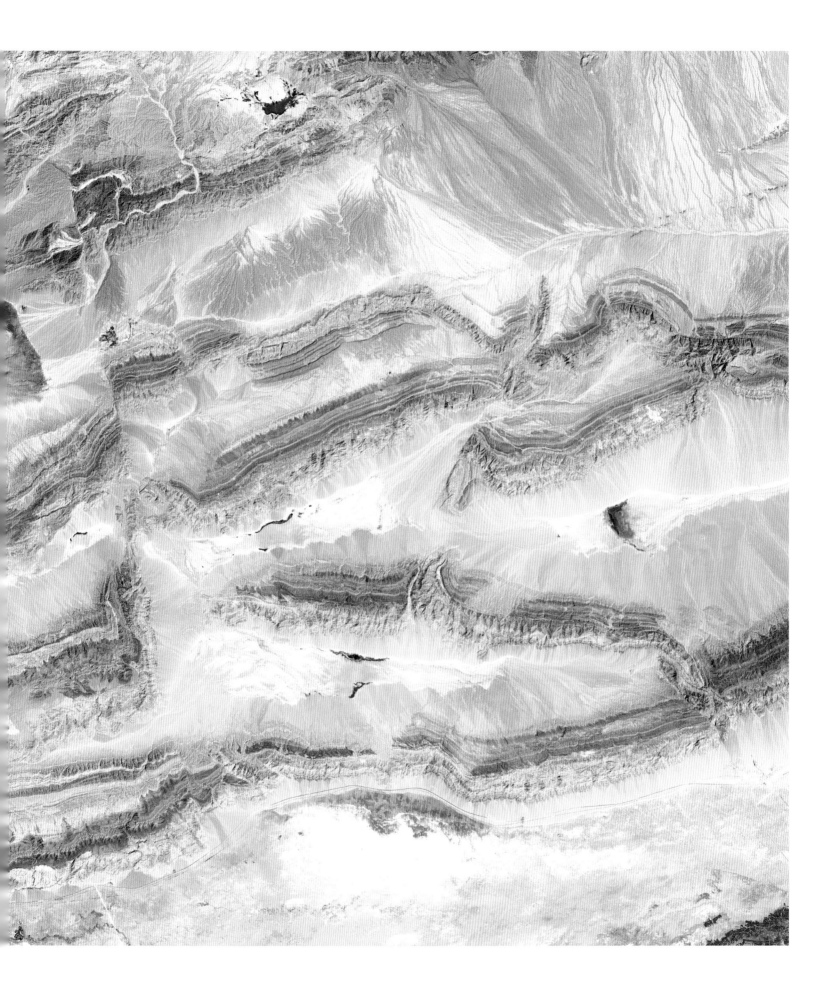

Aftermath of a landslide
A powerful earthquake that struck the island of Hokkaido, Japan, on 6 September 2018 triggered a series of devastating landslides as a secondary effect, engulfing homes and bringing down power lines.

earthquakes
and tsunamis

Earthquakes are ground tremors caused by a sudden release of energy deep underground. Motion at the surface occurs along fault lines, which are usually located close to plate boundaries. An earthquake's strength and destructive power depends on the amount of energy it releases. The magnitude of an earthquake, expressed on the Richter or moment-magnitude scales, is a measure of this power. Underground movement continues after the initial earthquake, creating smaller quakes called aftershocks. Earthquakes on the seafloor, barely detectable on the surface, can cause giant sea waves that grow into huge walls of water, known as tsunamis, as they reach coastlines.

WHAT CAUSES EARTHQUAKES
Underground movement of rocks along a fault causes energy to be released as seismic waves. Rocks begin to rupture or shift at a point underground called the focus. The point on Earth's surface directly above the focus is called the epicentre. Seismic waves, called surface waves, radiate out from the epicentre along Earth's surface, often triggering landslides, tsunamis, and damage to built structures.

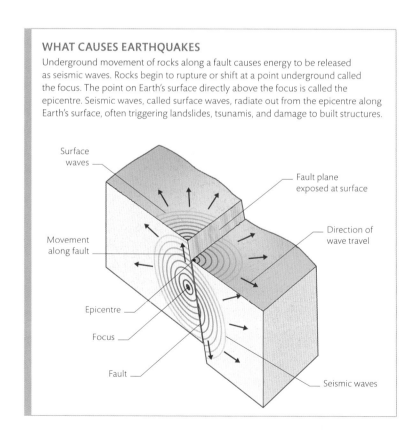

- Surface waves
- Fault plane exposed at surface
- Movement along fault
- Direction of wave travel
- Epicentre
- Focus
- Fault
- Seismic waves

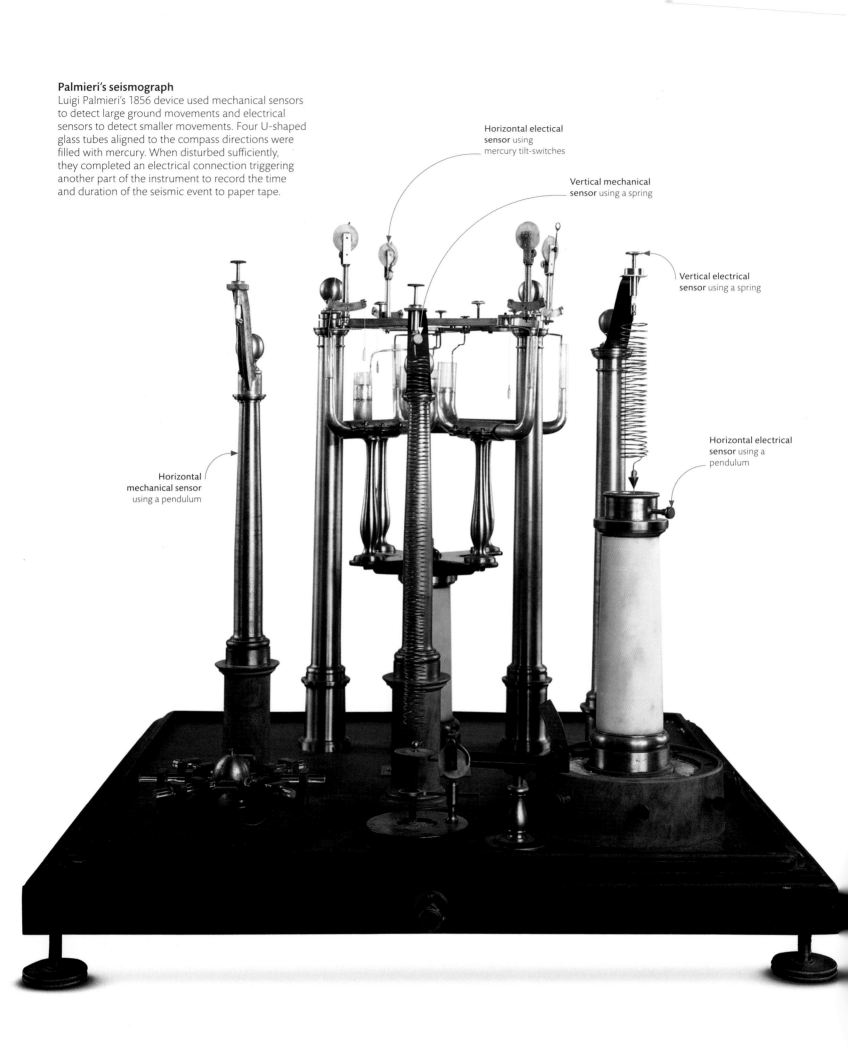

Palmieri's seismograph
Luigi Palmieri's 1856 device used mechanical sensors to detect large ground movements and electrical sensors to detect smaller movements. Four U-shaped glass tubes aligned to the compass directions were filled with mercury. When disturbed sufficiently, they completed an electrical connection triggering another part of the instrument to record the time and duration of the seismic event to paper tape.

Horizontal electical **sensor** using mercury tilt-switches

Vertical mechanical **sensor** using a spring

Vertical electrical **sensor** using a spring

Horizontal electrical **sensor** using a pendulum

Horizontal mechanical sensor using a pendulum

Tokyo earthquake on paper
A record produced by a seismometer is called a seismogram. This example, produced in Oxford, England, records the first disturbance from a major earthquake near Tokyo, Japan, shortly after 3 am on 1 September 1923.

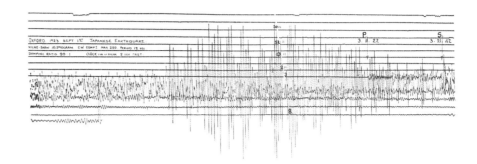

measuring earthquakes

Seismometers on the Moon
The most accurate seismometers are almost too sensitive for use on Earth due to the planet's constant seismic noise. But they were deployed on the Moon in 1969 by the Apollo 11 astronauts, where they detected tiny moonquakes and the impact of micrometeors from 1969 until 1977.

Ancient civilizations had a wide variety of explanations for the causes of earthquakes, from the activities of subterranean creatures to the presence of openings to the Underworld and the effects of winds below the surface. An accurate measurement of an earthquake's strength and duration was not possible until the invention of the seismometer.

The first known instrument for measuring the strength of earthquakes was invented by the Chinese mathematician Chang Heng in 132 CE. It is likely, but not certain, that Heng's device used a pendulum to detect the ground motion, as did European scientists investigating earthquakes in the 18th century.

Most of these early seismometers used the inertia of a weight, suspended on a pendulum or a spring, to provide a fixed reference point against which ground movement could be measured. While the ground and the instrument's framework shake, the weight remains relatively still. The changing distance between the frame and the weight measures the strength of the earthquake.

One of the first devices that could create a record of seismic events, known as a seismograph, was built by Luigi Palmieri at the Vesuvius Observatory in 1856. It recorded earthquakes that preceded the volcanic eruptions of Mount Vesuvius in 1861, 1868, and 1872 on a roll of paper.

Palmieri's seismograph was also the first to use electromagnetism to detect the motion. Instead of a spring or pendulum, modern seismometers use electronics to apply a magnetic or electrostatic force to keep the weight still. The most modern seismometers are sensitive enough to detect small, distant earthquakes, underground nuclear explosions, and even the tidal pull of the Moon on Earth's rocks.

Widely separated seismometers can be used to triangulate the location of seismic events. Since the 1980s, computer analysis of data from a global network of seismometers has been used to give a three-dimensional picture of Earth's deep interior, similar to the images provided by computerized tomography (CT) scans in medicine.

> " ... the seismograph ... enables us to see into the Earth and determine its nature. "

RICHARD DIXON OLDHAM, IRISH GEOLOGIST, 1906

MEASURING AND RECORDING EARTHQUAKES

A seismogram is a record of an earthquake's seismic waves, which are measured by an instrument called a seismograph. There are two types of seismic waves: body waves, which move from an earthquake's focus (the point underground where rocks start to rupture or shift) through Earth's interior; and surface waves, which travel from the epicentre (the point on Earth's surface located above the focus) along Earth's surface. The squiggly lines (amplitude) reflect the amount of energy released by the waves.

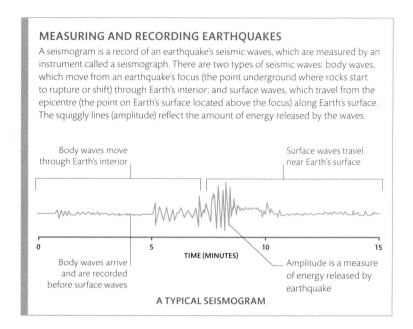

Body waves move through Earth's interior

Surface waves travel near Earth's surface

Body waves arrive and are recorded before surface waves

TIME (MINUTES)

Amplitude is a measure of energy released by earthquake

A TYPICAL SEISMOGRAM

shaking ground

Earthquakes are ground tremors caused by a sudden release of energy in underground rocks (see pp.128–29). Most last from a few seconds to a few minutes, but the vast amount of energy released can leave an imprint on a landscape for thousands of years. An earthquake's intensity at a given location is related to the distance from the fault, its depth, and the soil type. During all earthquakes, the ground starts to shake, but damage can come in many forms: large areas of rocks can move hundreds or thousands of kilometres along faults, landslides and mudflows can be triggered by seismic waves, and soil can liquify into flowing mud.

Rainbow patterns show deformation and ground movement

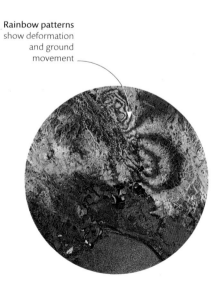

Ground-breaking damage
Earthquakes can cause immense damage to humans, landscapes, and infrastructure. The seismic waves that travel along Earth's surface are the ones that cause the most damage, often causing bridges and buildings to collapse, and roads (here, in Alaska, US, right) to fracture and break.

Earthquake by satellite
In 2014, the region of Napa Valley in California, US, experienced its largest earthquake for 25 years. The colours on this image depict radar signals acquired by satellite during the event.

Dome-like structure
flattened and lowered by
erosion and weathering

Vast tower made up of
individual, irregular-shaped
columns, about 30 m (100 ft)
high and 3 m (10 ft) wide

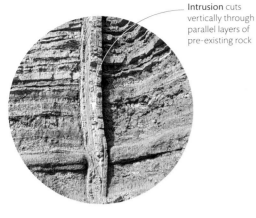

Magmatic dike
This dike was formed when hot, liquid magma pushed up through cracks in sedimentary rocks in Nahal Ardon, Makhtesh Ramon, Israel. The magma then cooled and solidified.

Intrusion cuts vertically through parallel layers of pre-existing rock

igneous intrusions

As molten magma rises from Earth's mantle through the crust, it squeezes through cracks in pre-existing rocks, widening them and sometimes forcing overlying rocks to bulge upwards. When the magma cools and solidifies beneath Earth's surface (rather than erupting through the surface as lava via fissures and volcanoes), it forms bodies of igneous rock called intrusions. If the overlying rocks become eroded, the intrusions can become exposed above the surface. Many of the world's mountain chains – such as the White Mountains in New Hampshire and the Yosemite Valley (see pp.68–69) in the Sierra Nevada in California, US – are made up of igneous intrusions that were exposed after the overlying rocks were eroded.

Devil's Tower
This dramatic geological feature (also known by its Lakota name Mato Tipila, meaning "bear lodge") rises above ponderosa pines in Wyoming, US. The tower is an intrusion formed from magma that cooled inside a magma chamber. It was once over 1.5 km (1 mile) below Earth's surface but was later exposed by erosion and lowering of the surrounding landscape. Made of an igneous rock called phonolite porphyry, it is the largest structure of its kind in the world.

Base of intrusion is blocky and irregular

TYPES OF IGNEOUS INTRUSION

There are several types of igneous intrusion. Dikes and sills are both slab-shaped intrusions that form in layers of pre-existing sedimentary rock called country rock. Dikes cut across the rock layers, while sills form parallel to them. Large, irregular-shaped intrusions are called plutons. The largest of these – batholiths – extend at least 100 square km (40 square miles) into the rock layers. The heat released by igneous intrusions can cause the surrounding rocks to undergo a process of chemical change called contact metamorphism (see pp.90–91).

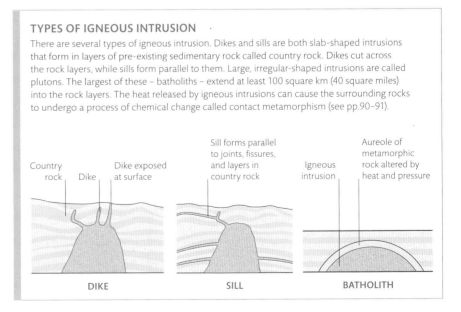

Country rock | Dike | Dike exposed at surface

Sill forms parallel to joints, fissures, and layers in country rock

Igneous intrusion

Aureole of metamorphic rock altered by heat and pressure

DIKE | SILL | BATHOLITH

Like a square-rigger ship sailing across a dry sea, Shiprock rises above the high desert of northwestern New Mexico, US, trailing a series of low ridges in its wake. The ridges are volcanic dikes – thin walls of resistant rock radiating away from the central peak, which is the throat of a long-dead volcano. The rock is fragmented, but nevertheless hard, volcanic breccia – fine debris from a series of explosions triggered when magma

Shiprock

pushed up through fractures in Earth's crust and interacted with groundwater beneath the surface. Dating studies suggest that the rocks of the formation solidified about 30–25 million years ago. Since then, erosion has taken away as much as 900 m (3,000 ft) of the overlying sandstones and shales, exposing the volcano's internal plumbing.

The composition of the dike rock suggests it was formed from magma that melted at great depth. Similar rocks have been found in igneous intrusions and surface flows elsewhere in New Mexico and Arizona, including in the distinctive rock formations of Monument Valley. These intrusions are the result of the tectonic forces that drove the uplift of the Colorado Plateau, which extends across much of Utah, New Mexico, Arizona, and Colorado.

Standing 482 m (1,583 ft) above the surrounding plain, Shiprock is a tempting climb. But the rock formation is sacred to the Navajo people, and climbing it has been banned since the 1970s. Where European settlers saw a sailing ship, the Navajo had always seen a "rock with wings", representative of the legendary great bird that brought the Navajo people to their land.

View from the ground

Pieces of breccia are scattered along one of the dikes

Aerial view
Two prominent dikes join to the jagged central peak of Shiprock, which is a volcanic plug – a mass of rock that once filled the vent of an active volcano. Five smaller dikes can also be traced across the immediate area. Together, they reveal the internal workings of the now extinct volcano.

Lake has formed inside the volcano's caldera

Mount Pinatubo

This stratovolcano (see panel, right) is situated on the eastern Kamchatka Peninsula, Russia. A lake of hot, acidic water has formed in the volcano's Troitsky Crater, which formed during an explosive eruption about 400 years ago.

Volcanic cones and craters

Dark grey basalt and red-rimmed craters give these low-lying volcanic cones and the surrounding landscape in Iceland an otherworldly appearance. The red colour of the deposits on the rims of the craters is produced by oxidation of iron minerals.

VOLCANO SHAPES

There are several types of volcano. Lava domes (also called volcanic domes) form when viscous lava accumulates over and around the vent and solidifies, forming a dome that can explode violently. Cinder cones (also known as pyroclastic cones) are formed mainly from fragments of gas-charged magma that is ejected from the vent as lava, then solidifies and falls as cinders. Shield volcanoes have gently sloping sides produced by thin, fluid lava that flows without an explosive eruption. Stratovolcanoes (also called composite cone volcanoes) have steep sides made from many layers of lava and other rocks fragments ejected in explosive eruptions.

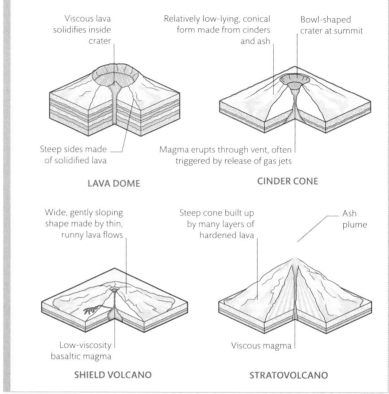

Viscous lava solidifies inside crater

Relatively low-lying, conical form made from cinders and ash

Bowl-shaped crater at summit

Steep sides made of solidified lava

Magma erupts through vent, often triggered by release of gas jets

LAVA DOME

CINDER CONE

Wide, gently sloping shape made by thin, runny lava flows

Steep cone built up by many layers of hardened lava

Ash plume

Low-viscosity basaltic magma

Viscous magma

SHIELD VOLCANO

STRATOVOLCANO

volcanoes

Volcanoes are landforms created when magma escapes from Earth's interior and reaches the surface through openings, or vents. They generally form at the boundaries of tectonic plates (see pp.110–11) or over mantle plumes – columns of extremely hot rock rising through the mantle and crust – and are found not only exposed on Earth's surface but also on the ocean floor and under icecaps. Once the magma erupts as lava, it cools and solidifies around the vents. Volcanoes consist of two main components: a volcanic cone, which is a mountain created by the accumulation of magma around the vent from successive eruptions; and a crater – a steep-walled, bowl-shaped depression surrounding the vent, from which magma erupts. If eruptions empty out all the magma held inside the reservoir (magma chamber) under the volcano, the cone may collapse, forming a circular depression called a caldera. There are various types of volcano; their shapes are determined largely by the properties of the magma and the way in which the volcano erupts (see p.141 and panel, above).

Tungurahua Volcano
Situated in Ecuador, Tungurahua (right) is a stratovolcano (see p.139) that erupts periodically in a Vulcanian or Strombolian style (see panel, below).

Vulcanian eruption
This dramatic Vulcanian eruption took place at the Fuego-Acatenango Massif – a series of volcanic vents in Guatemala. A large, dense plume of ash is ballooning over the volcano, while a pyroclastic flow hurtles down its flanks.

Relatively small **plume** of ash and steam rises from summit crater

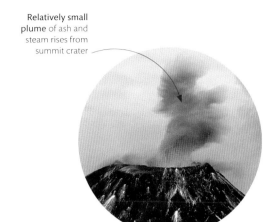

volcanic eruptions

Magma rising from Earth's mantle escapes to Earth's surface as lava through volcanoes during events called volcanic eruptions. These events can be very destructive or relatively calm, depending on the magma's chemistry and gas content. Magmas that contain large amounts of gas and are viscous (resistant to flow) tend to erupt explosively, with plumes of ash propelled into the atmosphere and pyroclastic flows – avalanches of ash, hot gases, and rock fragments that roll down the flanks of the volcano. Runny magmas with a low gas content tend to erupt as quiet, oozing lava flows. Explosive eruptions tend to be associated with steep-sided volcanoes, while quiet eruptions are associated with flat-lying, broad landforms (see pp.138–39).

TYPES OF VOLCANIC ERUPTION

There are several types of volcanic eruption, although a volcano can erupt in a number of ways and a single eruption can include elements from each type. Strombolian eruptions are associated with lava bombs and small to no ash clouds. In a fissure eruption, runny lava flows from a linear fissure, solidifying into low-lying landscapes. In Plinian eruptions, a tall gas and ash plume rises and falls on the flanks, accompanied by loud explosions. Vulcanian eruptions are short-lived, violent, intermittent events, accompanied by dense ash and gas clouds and, often, lava bombs.

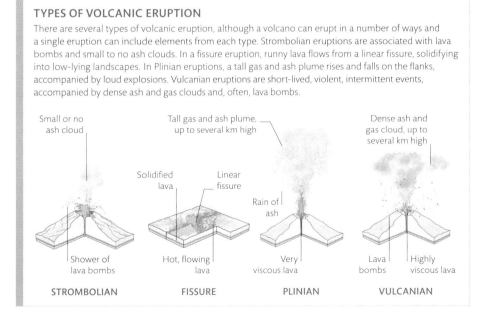

Small or no ash cloud

Tall gas and ash plume, up to several km high

Dense ash and gas cloud, up to several km high

Solidified lava

Linear fissure

Rain of ash

Shower of lava bombs

Hot, flowing lava

Very viscous lava

Lava bombs

Highly viscous lava

STROMBOLIAN   FISSURE   PLINIAN   VULCANIAN

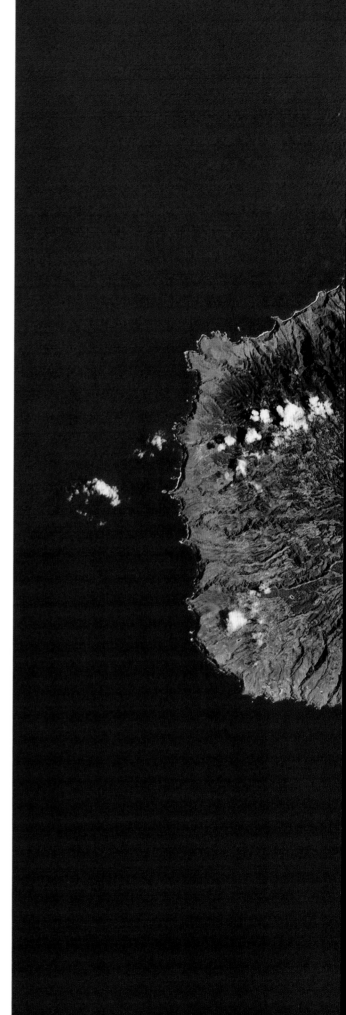

Hotspot volcano
La Palma, one of the Canary islands,
is a volcanic island formed over a hotspot.
In 2021 the Cumbre Vieja volcano erupted,
causing lava to flow down the mountainside.

hotspots

When a column of hot mantle material, called a plume, rises
to the base of the lithosphere, it creates a zone called a hotspot.
In these locations, the magma may be hot enough to melt the
overlying lithosphere and force its way to the surface to form
a volcano. About 100 so-called "hotspot volcanoes" are known.
Most of these are located within a tectonic plate rather than
at the boundaries between adjacent plates, but a few appear
on mid-ocean ridges, as in Iceland. Some hotspots lie within
continents – for example, under Yellowstone National Park, US.
Others underlie ocean basins, giving rise to volcanic island groups
such as the Canary Islands and Hawaii.

FORMATION OF A VOLCANIC ISLAND CHAIN

When hotspots underlie oceanic plates, they form volcanoes on the seafloor above.
As the tectonic plate moves over a hotspot, a chain of extinct volcanoes, or hotspot
track, forms. A good example can be seen in Hawaii today. Volcanic eruptions occur
only on the Big Island of Hawaii, the youngest island in the chain, which has three
active volcanoes. The islands to the northwest are the remains of ancient volcanoes.

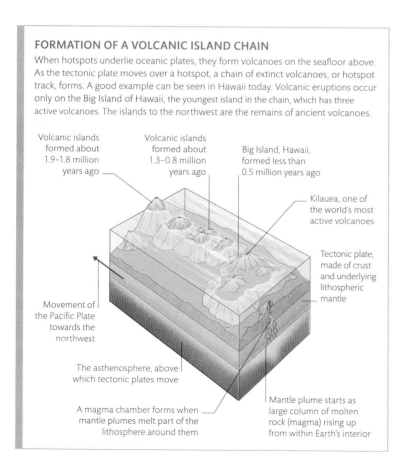

Volcanic islands
formed about
1.9–1.8 million
years ago

Volcanic islands
formed about
1.3–0.8 million
years ago

Big Island, Hawaii,
formed less than
0.5 million years ago

Kilauea, one of
the world's most
active volcanoes

Tectonic plate,
made of crust
and underlying
lithospheric
mantle

Movement of
the Pacific Plate
towards the
northwest

The asthenosphere, above
which tectonic plates move

A magma chamber forms when
mantle plumes melt part of the
lithosphere around them

Mantle plume starts as
large column of molten
rock (magma) rising up
from within Earth's interior

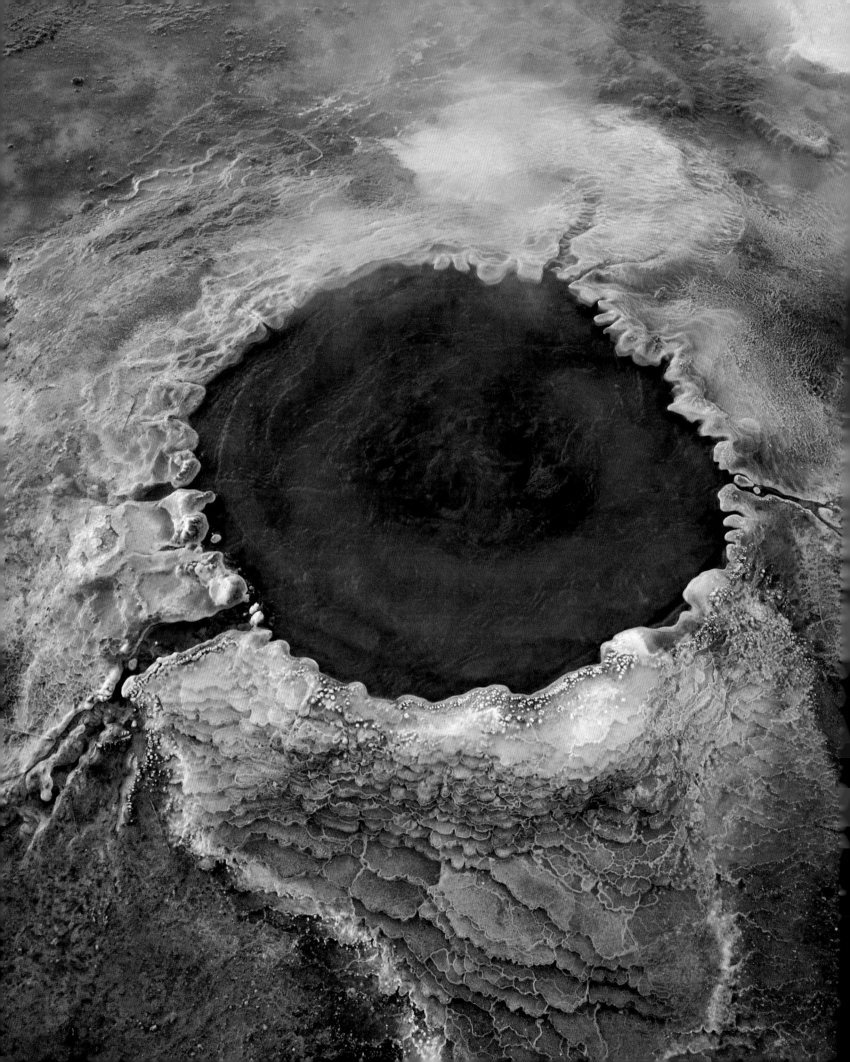

Colourful deposits
Some hot springs – such as this one in El Tatio,
Atacama, Chile – form near active volcanoes.
At the spring's surface, the hot water evaporates,
leaving behind mineral deposits called sinters.

geothermal
features

Geothermal activity provides some of the most striking
evidence of Earth's processes, many of which are driven by
its internal heat. The temperature of Earth rises with increasing
depth, from the cooler crust towards the hotter mantle and
core. A steep rise in temperature within a relatively shallow
depth often leads to activity at the surface that can take dramatic
forms. This occurs when groundwater seeps down into the
subsurface through fractures and is heated by warm rocks
or nearby magma. A combination of hot water and steam
rises back to Earth's surface through geothermal features such
as geysers (intermittent hot water spouts), hot springs (pools
of hot water), and fumaroles (columns of hot gas and steam).

HOW GEOTHERMAL FEATURES FORM

Groundwater enters the subsurface as part of the water cycle. Nearby volcanoes
or hot rocks can superheat the water to over 180°C (360°F). Hot springs form when
the water rises back to the surface and forms a pool. If water and steam become
trapped in hollows underground, pressure builds up until they are periodically
released in eruptions called geysers. If hot water turns into steam before reaching
the surface, steam and other gases are ejected through vents called fumaroles.

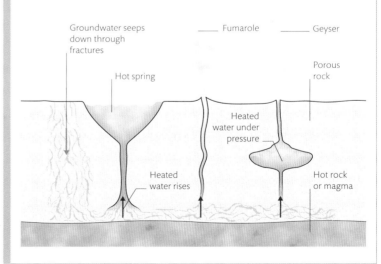

Groundwater seeps
down through
fractures

Fumarole

Geyser

Hot spring

Porous
rock

Heated
water under
pressure

Heated
water rises

Hot rock
or magma

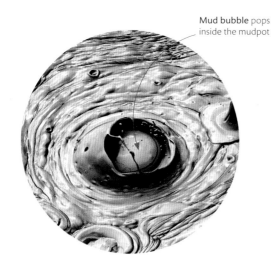

Mud bubble pops
inside the mudpot

Bubbling mudpot
Groundwater rising through Earth absorbs
gases that make it acidic. The acid in the
water then breaks down rock, which turns
into a pool of mud that bubbles and boils
at Earth's surface. This image (left) was
taken in Wai-o-tapu, New Zealand.

Fountain geyser
Iceland's Strokkur Geyser erupts every
few minutes, releasing hot groundwater
and steam in fountains about 30 m
(100 ft) high. Silica-rich mineral
deposits known as geyserites have
precipitated around the edge of the
water pool.

hot springs

Water constantly moves between Earth's surface and underground as part
of the global water cycle. Groundwater seeping down through cracks in Earth's
crust can become superheated (see pp.144–45) and then return to the surface
in the form of hot springs. These range from slow seeps to flows of more than
150 litres (40 gallons) per second. As the heated water rises, it may dissolve
minerals in the surrounding rock, which then precipitate out as solid, often
vibrantly coloured deposits when the water cools. Hot water can also erupt
as intermittent steam and hot-water spouts called geysers (see panel, below),
of which there are about 1,000 worldwide. Hot springs and geysers typically
occur at the boundaries of tectonic plates (see pp.110–13) or near volcanoes.

HOW A GEYSER ERUPTS

A geyser occurs where superheated water flows through an underground system of tubes and
chambers. A constriction in one of the tubes leads to an increase in pressure, which eventually
results in a blast of hot water erupting at Earth's surface. As the water pressure drops, the water
instantaneously turns to steam at the surface. After an eruption, groundwater percolates through
cracks into the chamber, recharging it, and the process starts again. Deposits called geyserites left
behind by mineral-rich waters line the chamber.

Geyserite-lined
underground
chamber

Steam circulating
at high pressure

Steam expands as
water pressure drops

Geyser
plume

Super-
heated
water

Constriction

Feeder channel
to surface

Geyser
outlet

Bedrock

Rising water

1. PRESSURE BUILD-UP

2. PRESSURE RELEASE

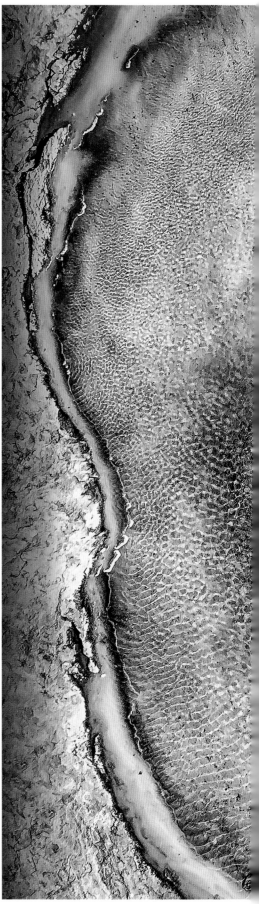

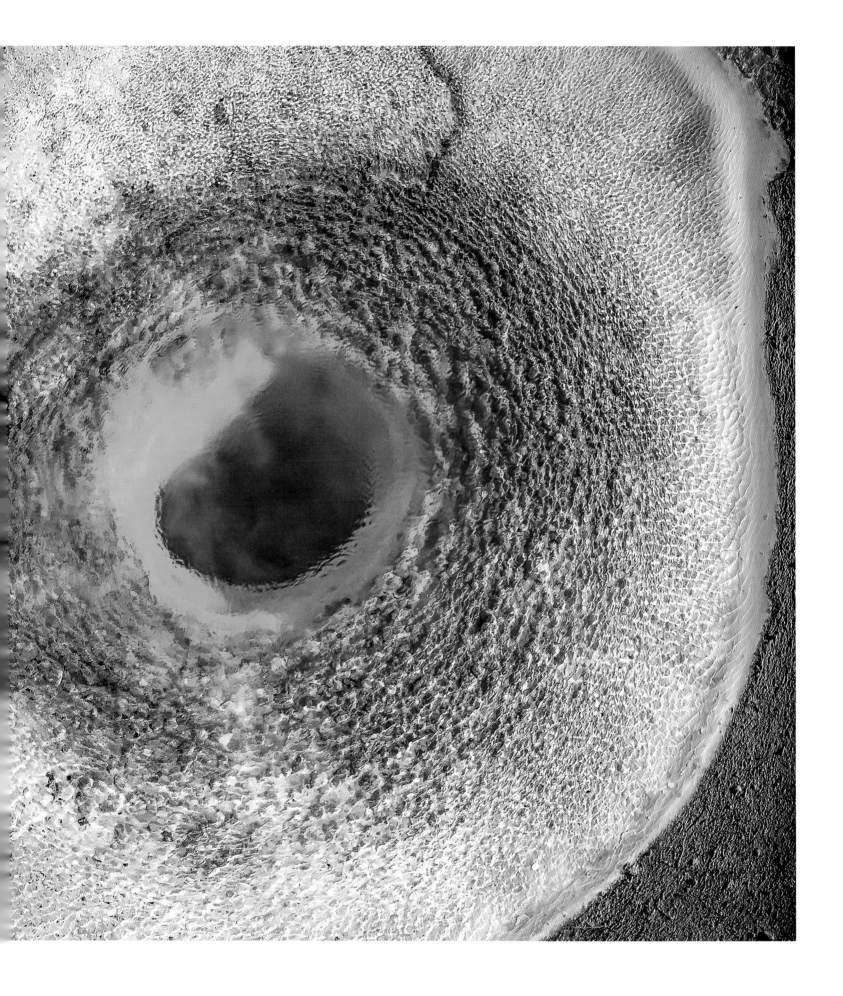

Potholes in rivers
Circular or cylindrical holes in the riverbed, called potholes, form when rock fragments carried by the river (here, Blyde River in South Africa, right) reach small depressions in the bedrock. The fragments whirl around the depression, eroding and scouring it.

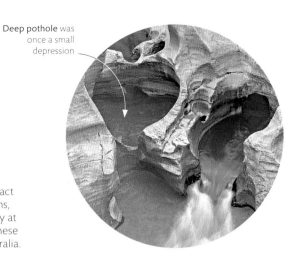

Deep pothole was once a small depression

Erosion from wave action
Coastlines are shaped by the impact of waves. Erosion cuts indentations, or notches, in coastal cliffs, usually at the level of high tide, as seen in these sandstone cliffs in Tasmania, Australia.

erosion by water

The water in oceans and rivers shapes the landscape. The repeated action of waves wears away coastlines, making them progressively straighter, and wave-based erosion of headlands – cliffs that project out into the sea – can carve out caves, arches, and sea stacks. As water from a river moves along its course, it erodes the rocks on which it flows, carrying sediments and rocks downstream and depositing them when the current slows. When the water is fast flowing – such as near the river mouth or in times of flood – it transports a greater volume of sediment.

RIVER TRANSPORT
Material eroded by water can be transported through several mechanisms, depending on the size of its particles. Fine-grained particles, such as silt and clay, are light enough to remain suspended in the water column while being carried along, while larger items such as pebbles are heavier and bounce along the riverbed – a process known as saltation. If the river is fast flowing, larger rocks and boulders can be dragged along the riverbed in a process called traction.

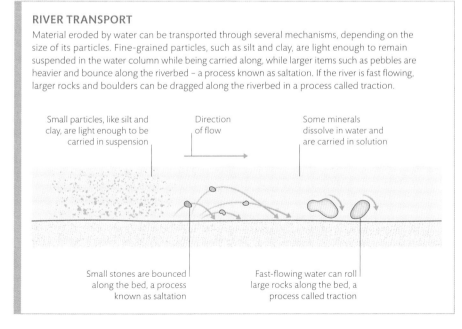

Small particles, like silt and clay, are light enough to be carried in suspension

Direction of flow

Some minerals dissolve in water and are carried in solution

Small stones are bounced along the bed, a process known as saltation

Fast-flowing water can roll large rocks along the bed, a process called traction

DESERT PAVEMENTS

Wind erosion in dry conditions can lead to the formation of a hard, stony surface called a desert pavement. This feature forms when fine-grained particles such as sand are removed from the desert's surface by wind in a process called deflation. As wind removes the finer particles, the coarse particles left behind become increasingly closely packed and concentrated. The land surface also lowers due to the particles' removal.

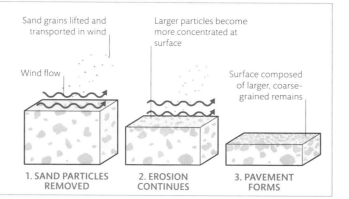

Sand grains lifted and transported in wind

Wind flow

Larger particles become more concentrated at surface

Surface composed of larger, coarse-grained remains

1. SAND PARTICLES REMOVED

2. EROSION CONTINUES

3. PAVEMENT FORMS

Carved by wind

In desert landscapes, eroded sand particles carried by winds collide with rock formations, sand-blasting them in a process called wind abrasion. Over time, this carves the rocks into smooth shapes, as seen in this image from the United Arab Emirates.

Tower-like structure carved into contorted shape by wind abrasion

Small gully formed by wind-blown particles eroding sandstone

erosion by wind

Wind is a powerful sculptor. It can remove and transport large quantities of sand, soil, and dust particles. In the process, it shapes Earth's surface – particularly in desert environments, which are covered with loose material, and in areas with little vegetation, since plants stabilize the soil and act as windbreaks. In deserts, wind gusts remove sand- and silt-sized particles, transporting them through the air and depositing them to form sand dunes, loess plateaus, and – over time – sedimentary rocks (see pp.74–77). The heavier particles, such as pebbles and cobbles, are left behind. Wind (aeolian) erosion can be caused by a light wind that rolls particles along the ground or a strong wind that lifts large numbers of small particles into the air to create dust storms.

Cloud of sand stretches over the Red Sea

Satellite image of a sandstorm
When thunderstorms or cyclones occur, strong winds can lift large amounts of sand and dust from bare, dry soils and transport them hundreds to thousands of kilometres.

Different bedding layers are visible in sandstone

U-shaped glaciated valley
This valley (left) in Mount Rainier National Park, Washington State, US, was carved into a U-shape by glaciers, which are still present on the peaks, and later modified in its lower sections by rivers.

Glacial scratches
These long scratches, called striations, were carved out by debris carried by glacial ice as it moved over the rocks.

Elongated, parallel grooves indicate direction in which glacier was moving

erosion by ice

Glaciers are large, slow-moving bodies of ice. As they move – usually down a slope or mountain valley – they carve and modify the landscape. There are two processes involved in glacial erosion: quarrying, in which ice dislodges and transports large rock fragments; and abrasion, when rock fragments embedded in ice scratch and polish underlying rocks. Several landforms are created during erosion, including cirques (bowl-shaped, steep-sided hollows in a mountain), horns (angular peaks formed from several glaciers that have eroded a mountain in different directions), arêtes (narrow ridges that form when two glaciers move down opposite sides of a mountain), and roche moutonnées (see panel, below).

SCRATCHING AND ROUNDING

Ice sheets and glaciers leave evidence of their movement in the forms of scratched and rounded rocks. Outcrops in the path of moving ice are often rounded on the upstream side through abrasion, but on the downstream side cracks and joints indicate rocks have been ripped away (or quarried). The result is an asymmetrical feature called a roche moutonnée.

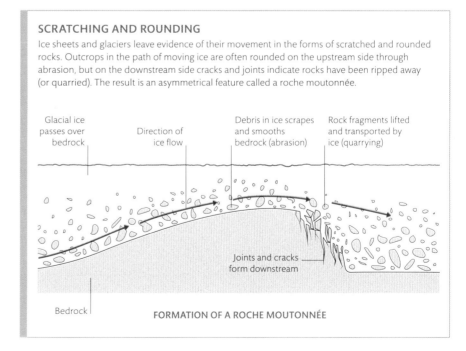

Glacial ice passes over bedrock

Direction of ice flow

Debris in ice scrapes and smooths bedrock (abrasion)

Rock fragments lifted and transported by ice (quarrying)

Joints and cracks form downstream

Bedrock

FORMATION OF A ROCHE MOUTONNÉE

weathering

The actions of water, wind, and temperature extremes can cause deterioration of rocks and minerals by breaking them down into smaller pieces or dissolving them. This process, called weathering, often acts in conjunction with erosion – weathering disintegrates or alters the rocks in situ, then erosion transports the disintegrated fragments away from the site. There are two main types of weathering: mechanical weathering (also called physical weathering), which involves the breaking down of rocks and minerals without altering their chemical composition; and chemical weathering, which changes their molecular structure. Typically, both types affect rock formations at the same time.

Towering pinnacles of weathered limestone

The Stone Forest
Groundwater and rain have chemically weathered this landscape in China's Yunnan Province to both sharp and rounded stone pillars. This type of weathering is typical of karst landscapes (see pp.166–67).

Frost-shattered **boulder** surrounded by scree, which typically forms a sloping heap

Sharp spikes produced by repeated frost-shattering

FROST-SHATTERING

Also called freeze-thaw weathering, frost-shattering is a form of mechanical weathering that occurs when water seeps into small cracks in a rock. When the water freezes, it expands by almost one-tenth of its volume, exerting pressure on the rock and enlarging the cracks, which allows more water to enter the rock. If a cycle of freezing and thawing is established, the rock will eventually break up. This type of weathering is typical of environments where surface water is abundant and temperatures repeatedly fluctuate around 0°C (32°F).

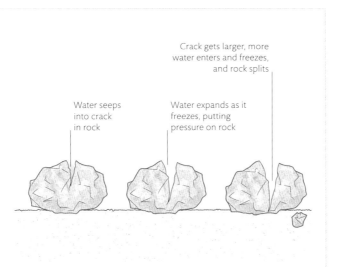

Water seeps into crack in rock

Water expands as it freezes, putting pressure on rock

Crack gets larger, more water enters and freezes, and rock splits

Castle of the Winds (Castell y Gwynt)
This rock formation, situated near the summit of Glyder Fach, in Snowdonia's National Park, Wales, consists of volcanic rocks from the Ordovician period (about 450 million years ago). The landscape and rocks were shaped and eroded by glaciers (see pp.152–53) in the last ice age, and broken down by frost-shattering (see panel, opposite), leaving behind tors (prominent rock outcrops) and scree (piles of stones).

Mountainous landscape smoothened by glaciers in last ice age

depositing sediment

Moving wind, water, and ice can all pick up and move particles of sediment. When the energy of movement drops – for example, when a river reaches a flatter section of its course and slows down – the effect of gravity means that the particles of sediment start to drop out of the flow, with the densest particles falling out first. Sediments often accumulate in layers in natural traps such as lake basins. If these sediments remain undisturbed, they may become compacted and cemented to form sedimentary rock, often with the original layers intact. Water, and particularly rivers, play a major role in sediment and transport and deposition (see pp.158–59). The sediments they deposit form features ranging from sand bars and gravel bars to alluvial fans and deltas (see pp.162–63).

Pebbles and cobbles
deposited by muddy
glacial meltwater

Alluvial fan
This alluvial fan formed when a river flowing from the Kunlun and Altun mountains flooded an open plain in the Taklamakan Desert, China. During this rare flood, the river deposited its sediment in the desert. The area soon dried out again, leaving this intricate pattern of river channels and sand bars behind. The area shown in this satellite image is about 60 km (38 miles) wide. The white, green, and blue areas indicate differing levels of moisture.

Deposition by ice
The rocks and sediment carried by a glacier and deposited when the ice melts are known as moraine. This material may later be retransported and reshaped by streams formed from meltwater (see pp.174–75).

RIVER CURRENTS AND SEDIMENT DEPOSITION

There are several reasons why a river deposits its sediment. Usually, it is because the water loses energy, for example as the river widens and the flow of water spreads out and slows down. Larger, heavier particles, such as pebbles and sand, fall out of suspension and are deposited first, while the lighter, finer particles, such as silt and clay, fall when the water current has slowed further. After the fine sediment settles, it can accumulate to form a sand bar or an alluvial fan.

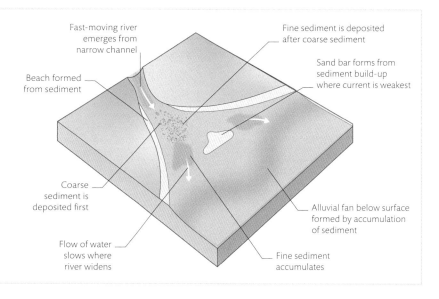

Fast-moving river emerges from narrow channel

Fine sediment is deposited after coarse sediment

Beach formed from sediment

Sand bar forms from sediment build-up where current is weakest

Coarse sediment is deposited first

Alluvial fan below surface formed by accumulation of sediment

Flow of water slows where river widens

Fine sediment accumulates

Water becomes shallower, with some rocks exposed above the surface

Meandering river
Streams called tributaries are joining this river (right) as it flows downstream in Iceland. Where the river bends, erosion has taken place on one side, resulting in the formation of river cliffs, while gravel is deposited on the other side (see panel, below).

Fast-flowing water
Sudden drops in gradient along a river's course can result in the formation of rapids – very fast-flowing, turbulent parts of a river.

rivers

A river is a large, natural stream of fresh water that starts on high ground, such as a mountain or a hill, then flows downstream along a defined course, or channel, to end its journey in a larger body of water – typically the ocean or sea, but sometimes a lake. A river's starting point is called its source, or headwater. The water can come from accumulated meltwater from glaciers and snow, from rainfall, or where groundwater flows up to the surface in the form of a spring. The river's end is called its mouth. A river shapes a landscape by eroding its bed and by transporting and then depositing sediment either downstream or across a wide area of its valley during a flood (see p.149 and panel, below).

HOW A MEANDER DEVELOPS
As rivers flow over relatively flat landscapes, they develop loops, or bends, called meanders. Because the velocity of the water is greater on the outer edge of a bend, the riverbed is more eroded there. Sediment is removed from the outside of the bend, sometimes forming a river cliff on the bank, and it is deposited downstream, in the inner bank of the next bend – an area called the slip-off slope. Over time, as increasing amounts of sediment are eroded and deposited, meanders become more pronounced. These features are called incised meanders.

Erosion on outside of bend

Sediment deposited on inside of bend, where flow is slowest

Curvature is accentuated as erosion and deposition continue

Line of fastest flow

River cliff

Slip-off slope

Narrowing neck of land on inside of bend

1. YOUNG MEANDER

2. INCISED MEANDER

Partially frozen waterfall
In winter, some of the water cascading in waterfalls can freeze solid. This waterfall in Iceland (right) is partially frozen and partially liquid. In the plunge pool below, ice and liquid water co-exist and mix.

Snow and icicles accumulate alongside cascading liquid water

Segmented waterfall
This broad waterfall in eastern Java, Indonesia, is cascading down a steep, amphitheatre-shaped ridge of rocks. Several frothy streams of different sizes are plummeting down the ridge and converging into a plunge pool below. When a waterfall consists of several, distinct streams it is called segmented.

waterfalls

Also called cascades, waterfalls form when a river or other body of water flows over a steep rock ledge and falls into a pool (called a plunge pool) below. Waterfalls can form in several ways but are usually caused by erosion in the upper stages of a river, where fast-flowing water carrying sediment flows from hard rock to softer rock (see panel, below). Waterfalls that form in this way usually retreat upstream, sometimes forming a gorge. Dramatic changes in a landscape – for example, caused by an earthquake, a volcanic eruption, or a landslide – can also lead to the formation of a waterfall. The shape of a waterfall depends on the underlying geology as well as the size and shape of the river.

HOW A WATERFALL FORMS
Waterfalls form where a river flows over a layer of hard rock (such as granite) onto an underlying layer of softer rock (such as limestone or sandstone). Since the water erodes the softer rock faster than the harder rock, the harder rock creates an overhanging ledge, called caprock. Eventually, the caprock collapses since it is no longer supported, and rocks fall into the plunge pool below, where they swirl around and contribute to further erosion of the softer rock.

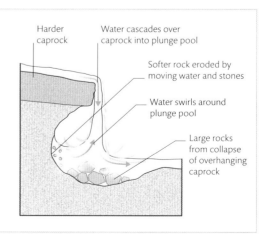

Harder caprock

Water cascades over caprock into plunge pool

Softer rock eroded by moving water and stones

Water swirls around plunge pool

Large rocks from collapse of overhanging caprock

ESTUARY ZONES

An estuary can be divided into zones, with each zone having different degrees of tidal influence and proportions of sea water to fresh water. At its furthest end, where the river meets the sea, the estuary is dominated by sea water. The middle of the estuary contains an almost equal mixture of sea water and fresh water. At the upper zone of an estuary, fresh water begins to encounter salt water driven by tides.

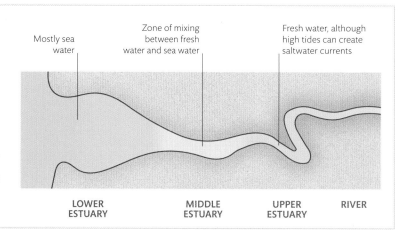

Mostly sea water

Zone of mixing between fresh water and sea water

Fresh water, although high tides can create saltwater currents

LOWER ESTUARY   MIDDLE ESTUARY   UPPER ESTUARY   RIVER

S-shaped river meanders through wetlands

Yukon Delta

The Yukon River starts its life in British Columbia, Canada, and flows through the Yukon Territory before entering Alaska and forming a delta in the Bering Sea. This satellite image shows the many meandering channels that make up the fan-shaped delta. Scientists believe that sediment flow into the delta is increasing due to climate-related thawing of permafrost in the basin, which is releasing more sediment held in ice.

Inland river delta

The Okavango River flows almost 1,000 km (600 miles) from Angola through Namibia and into a depression in Botswana. Unlike most of the world's rivers, the Okavango does not empty into a body of water; instead, it empties into a plain in the Kalahari Desert, where it ultimately evaporates or is transpired.

deltas and estuaries

Deltas form where rivers empty their water and sediment into a larger body of water – typically the sea, but sometimes a lake or another river – or, very rarely, into land (see above). Since a river loses energy and velocity as it nears its mouth, sediment often falls and settles on the riverbed. If large amounts of sediment (mainly mud, silt, and sand) build up, the river divides into a number of smaller, shallower channels, and new land (the delta) is formed. If a river flows more slowly and carries insufficient sediment to form a delta, sea water may enter the river mouth, creating a channel or inlet of brackish (slightly salty) water called an estuary. Within an estuary, the water level and salinity are affected by the tides (see panel, above).

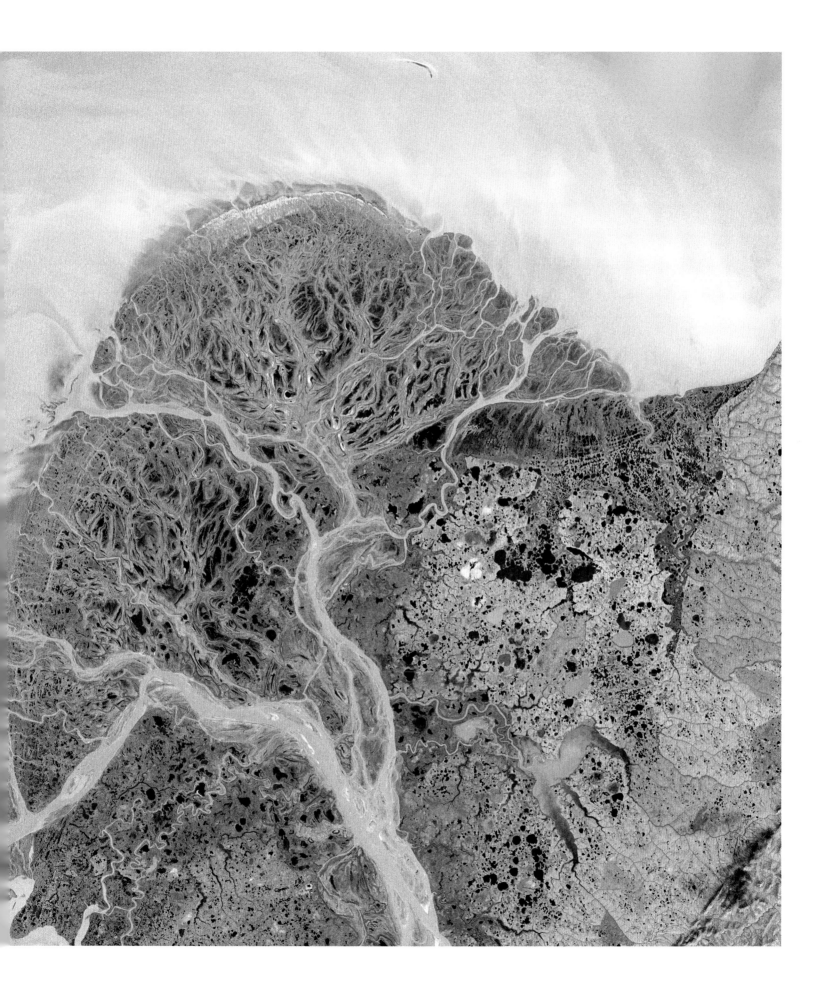

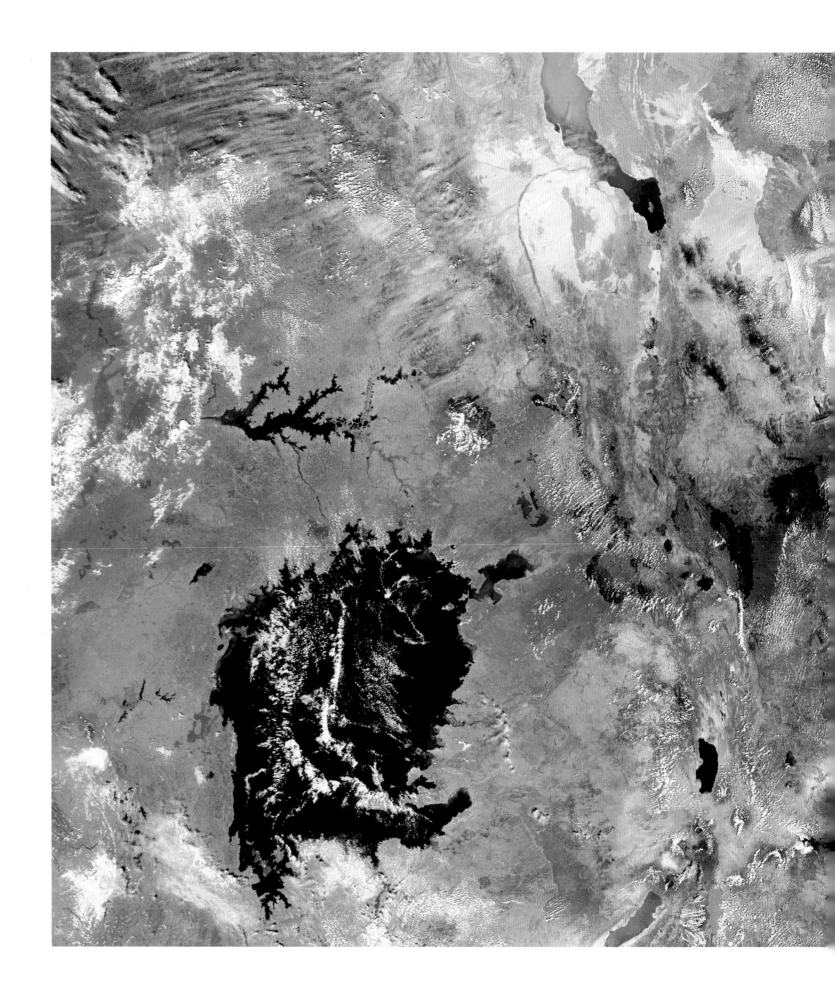

Lake Victoria
Located in East Africa's Rift Valley, Lake Victoria (captured here in a satellite image, bottom left) is a tectonic lake that formed when continental crust thinned close to a divergent boundary (see p.116). The largest lake in Africa, it lies mainly in Tanzania and Uganda but also extends into Kenya. The long, green lake at the top of the image is Kenya's Lake Turkana.

Lake Superior
Powerful storms blowing over the water's surface can create tall waves in large lakes, such as Lake Superior on the US–Canada border. The world's largest freshwater lake by surface area, Lake Superior has waves sometimes reaching several metres high.

Large, wind-driven wave crashes against lakeshore

lakes

Surrounded by land, and unconnected to the sea except by rivers or streams, lakes consist of a shallow depression called a basin that fills up with water from rain, melting snow and ice, rivers, or groundwater. Lake basins form in various ways. Most commonly, they are made by glaciers, which leave scoured depressions after they have retreated. Tectonic lakes form when movements of Earth's crust – such as at diverging plate boundaries (see pp.116–17) or along faults (see pp.126–27) – create a shallow depression. The water in lakes is relatively still compared to that of rivers, but it is still affected by currents driven by waves, wind, temperature changes, and river water entering the lake. Some lakes are entirely man-made.

TEMPERATURE LAYERS
In summer, many lakes have three distinct temperature zones: a relatively warm top layer (epilimnion); a colder bottom layer (hypolimnion); and a transition layer in between (thermocline), in which temperatures change rapidly. In autumn, seasonal mixing, called turnover, occurs, and the different layers combine until the entire lake is the same temperature. In winter, the layers become distinct again, although the temperatures are reversed: it is coldest at the icy surface and warmest on the lakebed. In spring, turnover re-occurs, during which the water is circulated from top to bottom. The warmer water is brought to the top, where the surface gains further heat from the Sun.

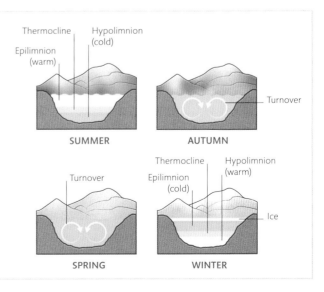

Thermocline · Hypolimnion (cold)
Epilimnion (warm)
SUMMER

Turnover
AUTUMN

Turnover
SPRING

Thermocline · Hypolimnion (warm)
Epilimnion (cold)
Ice
WINTER

Towers shaped by water
This karst landscape in China is the result of millions of years of erosion of carbonate rocks by water. This process left behind an array of towering pinnacles that are now topped by lush vegetation.

karst landscapes

From giant pinnacles to densely cracked rock pavements, karst landscapes contain some of Earth's most spectacular features. These landscapes are sculpted by the action of precipitation and groundwater, which is slightly acidic and slowly dissolves carbonate rock such as limestone or dolomite. First, cracks in the rock start to appear. The cracks widen both on the surface and underground, where a complex system of caves and streams develops at the level of the water table (the line below which rock is saturated with water). Mineral-rich groundwater drips into the caves, forming icicle-like deposits called stalactites and mounds on the cave floor called stalagmites. When cave roofs collapse, towering spires are left behind.

HOW KARST LANDSCAPES DEVELOP

When water flows over carbonate rock, it slowly erodes it, opening up cracks called grykes and hollows called kamenitzas. The blocks surrounded by the grykes are called clints. Water flowing through the cracks dissolves the rock further, opening up large underground cavities called caverns. As water continues to flow through the cavern, it becomes larger. Eventually, the roof of the cavern may become too thin and falls, leaving behind the walls, which form towers in the landscape. Vegetation becomes established in soil inside cracks in the rock.

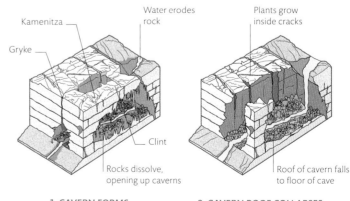

Kamenitza

Gryke

Water erodes rock

Plants grow inside cracks

Clint

Rocks dissolve, opening up caverns

Roof of cavern falls to floor of cave

1. CAVERN FORMS   **2. CAVERN ROOF COLLAPSES**

caves

Caves are underground cavities that typically form in karst landscapes (see pp.166–67). Over time, acidic groundwater and precipitation seep through cracks in limestone (or sometimes dolomite), enlarging the cracks until they widen to form underground passages. Many cave systems descend steeply to the water table and then continue sideways as flooded, tubular passages, with the water sometimes re-emerging at the surface as a spring. Eventually, if the water table drops, the caves dry out and sometimes the roofs collapse to form sinkholes. Dry caves can still widen as more precipitation and groundwater seep in. If the water table rises, dry caves can fill with water once again.

Stalactites hang from cave ceiling

Stalagmites form mounds on cave floor

Underwater cave

A cave diver explores a cenote cave system – an underground space filled with water that contains an opening to the outside. This cave system, in Yucatan, Mexico, was dry during the Last Glacial Maximum, about 22,000 years ago, when the ice sheets of the most recent ice ages were at their greatest extent and sea level was about 120 m (400 ft) lower than today.

Cave formations

This cave (left) is filled with limestone deposits that precipitate from water seeping into caves. Among the deposits are stalactites and stalagmites (see panel, below). Thin, hollow stalactites are called soda straws.

STALACTITES AND STALAGMITES

Rocks that form from mineral deposits inside caves are called speleothems. Two of the most common types are stalactites and stalagmites. As mineral-saturated water enters a cave's roof and drips down, residue builds up and stalactites form. As water droplets hit the ground, the minerals they deposit create mounds that grow into stalagmites. Over a long period of time, the two may join to form a single pillar that extends from floor to ceiling.

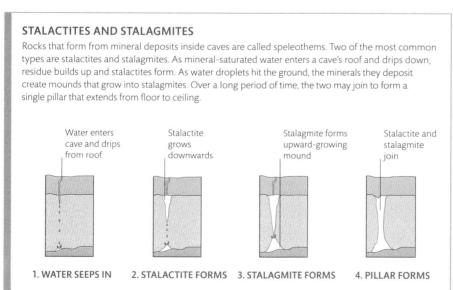

Water enters cave and drips from roof

Stalactite grows downwards

Stalagmite forms upward-growing mound

Stalactite and stalagmite join

1. WATER SEEPS IN   2. STALACTITE FORMS   3. STALAGMITE FORMS   4. PILLAR FORMS

Confluence of glaciers
At 12.4 km (7.7 miles) in length, the Gorner Glacier (left) is joined by the smaller Grentz Glacier (right) as it descends the western slope of Monte Rosa in the Valais Alps, Switzerland. It forms the second largest glacial system in the Alps. The dark stripe of the medial moraine is clearly visible.

glaciation

Glaciers are moving bodies of ice that form from the accumulation and compaction of snow over thousands of years. Often dating from the last ice age, they are present in many mountain ranges, especially in cold regions. As they move downslope due to gravity and the pressure of their own weight, glaciers erode the landscape to form valleys and in the process create deposits of rocks and sediments. Their abrasive action also leaves behind features such as striations, elongate scratches on rock surfaces, and grooves scooped into rocks in the landscape (see also pp.148–49).

GLACIAL EROSION AND DEPOSITION

Two glaciers, originating in semicircular depressions called cirques, may join up to form a single glacier that flows into a pre-existing valley. As they move, glaciers erode sediments by abrasion (when rock fragments on the underside scrape the surface of underlying rocks) and quarrying (when the glacier plucks out and incorporates bedrock from its base). The accumulated sediments left downslope are termed moraines. Terminal moraines are deposited at the lowest end of the glacier (snout), while lateral and medial moraines are deposited at its sides and in the centre.

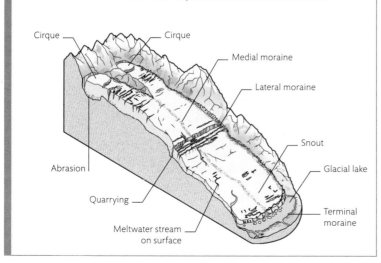

Cirque — Cirque
Medial moraine
Lateral moraine
Abrasion
Snout
Quarrying
Glacial lake
Meltwater stream
on surface
Terminal
moraine

Glacier Bay is an arm of the Pacific Ocean that thrusts deep into a spectacular landscape in southern Alaska. The bay is 105 km (65 miles) long and surrounded by some of the highest coastal mountains in the world, which were thrown up by the collision of the Pacific and North American tectonic plates. To the west, the Fairweather Range reaches 4,600 m (15,000 ft) above sea level. Its high peaks trigger snowfall from the moist

spotlight Glacier Bay

air blowing in from the Gulf of Alaska, feeding more than a thousand glaciers in the surrounding area.

Some of the glaciers flow directly into the sea as tidewater glaciers, while others terminate inland as valley glaciers. The Johns Hopkins and Margerie are two of the largest tidewater glaciers – each over a mile wide – with ice fronts up to 60 m (200 ft) high, where icebergs calve off into the water. Most of the region's glaciers are retreating as the climate warms. The Grand Pacific Glacier no longer reaches the ocean.

Glacier Bay was not always a bay. In 1680, during the Little Ice Age, its southern half was a broad valley occupied by the Tlingit people. A huge glacier was advancing from the north, and by 1750 the whole valley lay under thousands of feet of ice. But the surge was short-lived. By 1880, the ice had retreated 70 km (45 miles) inland, leaving behind Glacier Bay. Today, the tributary glaciers have retreated further up their valleys, which are now filled with water to form fiords.

The glacier is almost 1.5 km (1 mile) wide, 30 km (19 miles) long, and 50 m (165 ft) high at its terminus

The Lamplugh Glacier

Topeka Glacier
On the western side of Glacier Bay, the Topeka Glacier has carved a steep valley. Like 95 per cent of the glaciers in Alaska, it is thinning and now no longer reaches the sea. It is estimated that the area has lost 11 per cent of its glacial ice since the 1950s.

Glacier cave
Meltwater that runs through or under a glacier's ice can carve out a tunnel, or cave, between the ice and the bedrock. Unlike ice caves, which are formed in bedrock and contain ice all year round, glacier caves (here, in the Alps, left) are a seasonal feature. Increased melting due to climate change means glacier caves are becoming rarer.

Surface meltwater
This aerial view of the surface of Sawyer Glacier in Alaska, US, shows a meltwater stream entering a moulin – a vertical shaft in a glacier. The meltwater travels through the moulin to the glacier's bed.

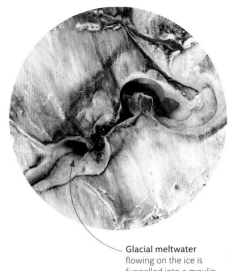

Glacial meltwater flowing on the ice is funnelled into a moulin

meltwater

When snow and ice held in ice sheets and glaciers melt, the liquid released is called meltwater. Streams of meltwater flow away from the glacier's terminus (the cliff-like edge where the glacier ends) and deposit debris – including rocks carried by the ice – downstream (see p.157). Meltwater streams can also create tunnels of water under a glacier, where they deposit sediment that can build up to form a long, sinuous ridge called an esker. Each year, glaciers melt naturally as part of the seasonal thaw. However, glaciers are melting at unprecedented rates worldwide as a result of climate change, contributing to a rise in sea level and increased flood risk.

MELTWATER LAKES
Lakes made of glacial meltwater sometimes form between the edge of a glacier and a moraine (a ridge of debris deposited by the glacier). When the melting glacier releases chunks of ice in a process called calving, the sudden input of material can cause waves, called seiche waves, on the lake's surface. The waves can move water over the moraine – a process known as overtopping – which may lead to flooding. Increased glacier melt due to global warming has led to an increase in size and number of meltwater lakes (also called proglacial lakes).

Glacier terminus

Chunks of ice break off

Lake forms from meltwater

Terminal moraine traps meltwater

Seiche wave can push water over moraine

Bedrock

ocean and atmosphere

On Earth's formation, its most volatile ingredients jetted out of its crust as gas. Some escaped into space, some condensed into oceans, and some coalesced to form the planet's outermost, gaseous layers – its atmosphere. These fluid parts of Earth remain in constant motion and interaction – the Sun's heat maintaining currents in the ocean and winds and weather systems in the air.

Aerial view of salt pans, France
Salt pans are shallow coastal ponds used to obtain salt from sea water. Sun and wind evaporate the sea water, concentrating the salts. Microscopic algae thrive in such extreme conditions. Different species bloom as salinity increases, changing the ponds' colour from light green to vibrant red.

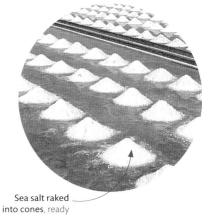

Sea salt raked into cones, ready for collection

Salt pan harvest, Thailand
When enough water has evaporated and salinity is sufficiently high, the salt crystallizes out of solution, forming crusts of salt that can be harvested.

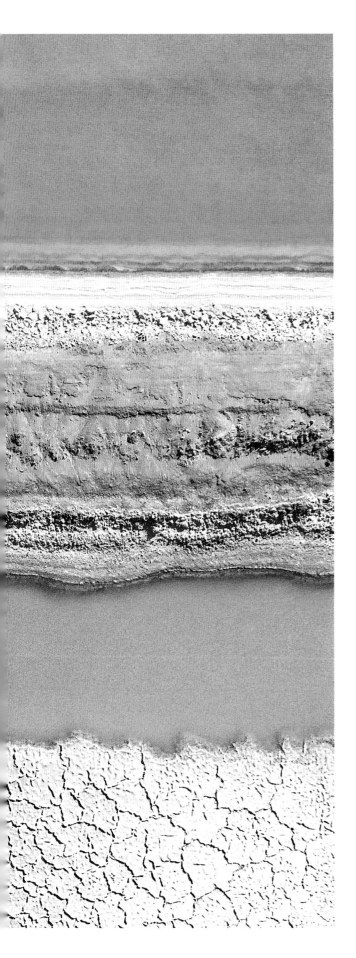

ocean chemistry

Sea water is a chemical cocktail of nearly 100 different elements, including life's building blocks – carbon, nitrogen, phosphorus, hydrogen, and oxygen – and even gold. Besides water (H_2O), the ocean contains 500 trillion tonnes of dissolved salts, the most common being chloride, sodium, sulfate, magnesium, calcium, potassium, and bicarbonate. The average concentration of salts (salinity) is 3.5 per cent. Some marginal seas, such as the Mediterranean and the Red Sea, are saltier; others, like the Baltic, are less so. Sea water is slightly alkaline (pH 7.5–8.4), and it holds 60 times more dissolved carbon dioxide than the atmosphere.

THE SALT CYCLE
Ocean chemistry is in constant flux, but overall salinity remains unchanged due to a balancing process called the salt cycle. Rivers, volcanoes, and deserts add around 3 billion tonnes of dissolved chemicals to seas annually, and a similar quantity enters the ocean at mid-ocean ridges and from the dissolution of seabed minerals. An equal amount is removed by living organisms and the deposition of their remains on the seafloor, and by the subduction of marine sediments at oceanic trenches.

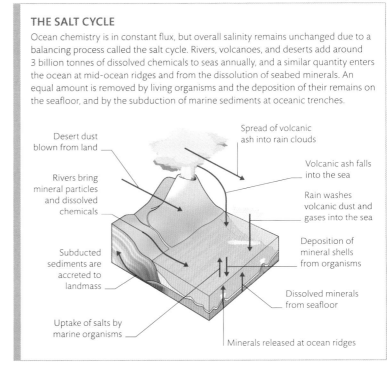

Desert dust blown from land

Spread of volcanic ash into rain clouds

Rivers bring mineral particles and dissolved chemicals

Volcanic ash falls into the sea

Rain washes volcanic dust and gases into the sea

Subducted sediments are accreted to landmass

Deposition of mineral shells from organisms

Dissolved minerals from seafloor

Uptake of salts by marine organisms

Minerals released at ocean ridges

Atlantic eddies
Based on satellite readings of ocean temperatures, this graphic reveals the currents that flow across the Atlantic. The Gulf Stream, indicated in yellow (warm) and green (cool), flows from Florida to northwest Europe. It is fed through the Caribbean by the warm Atlantic Equatorial Current (shown in orange and red). Both major current systems flow in a series of giant eddies.

ocean circulation

The constant turmoil of atmospheric winds, powered by solar energy, drives the world's major ocean currents. Surface circulation consists of a series of wind-driven gyres (circular currents) around each major ocean north and south of the equator (the North Atlantic gyre is pictured here). The huge volumes of water at the core of these gyres exert great pressure. This combines with the effects of Earth's rotation (the Coriolis effect), the position of land masses, and the narrow gateways between ocean basins, to keep ocean waters in perpetual motion, moderating climate throughout the world.

OVERTURNING CIRCULATION

As surface currents flow towards the poles, so bottom water, driven by changes in temperature and salinity, diffuses slowly back towards the equator (see p.185). This overturning circulation transports vast amounts of heat, dissolved chemicals, and sediments around the world's oceans. It acts as a regulator of Earth's climate and locks carbon dioxide in the oceans.

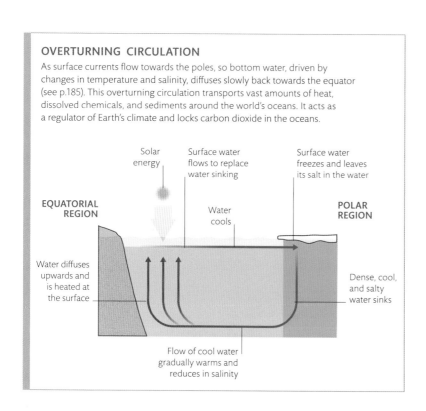

Solar energy

Surface water flows to replace water sinking

Surface water freezes and leaves its salt in the water

EQUATORIAL REGION

Water cools

POLAR REGION

Water diffuses upwards and is heated at the surface

Dense, cool, and salty water sinks

Flow of cool water gradually warms and reduces in salinity

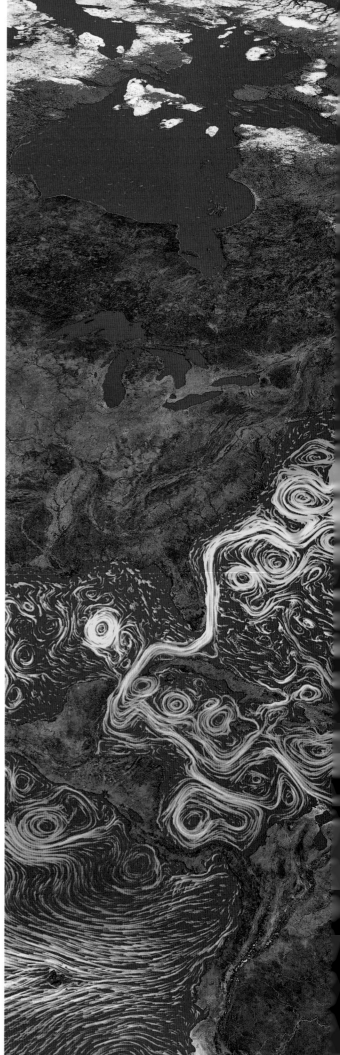

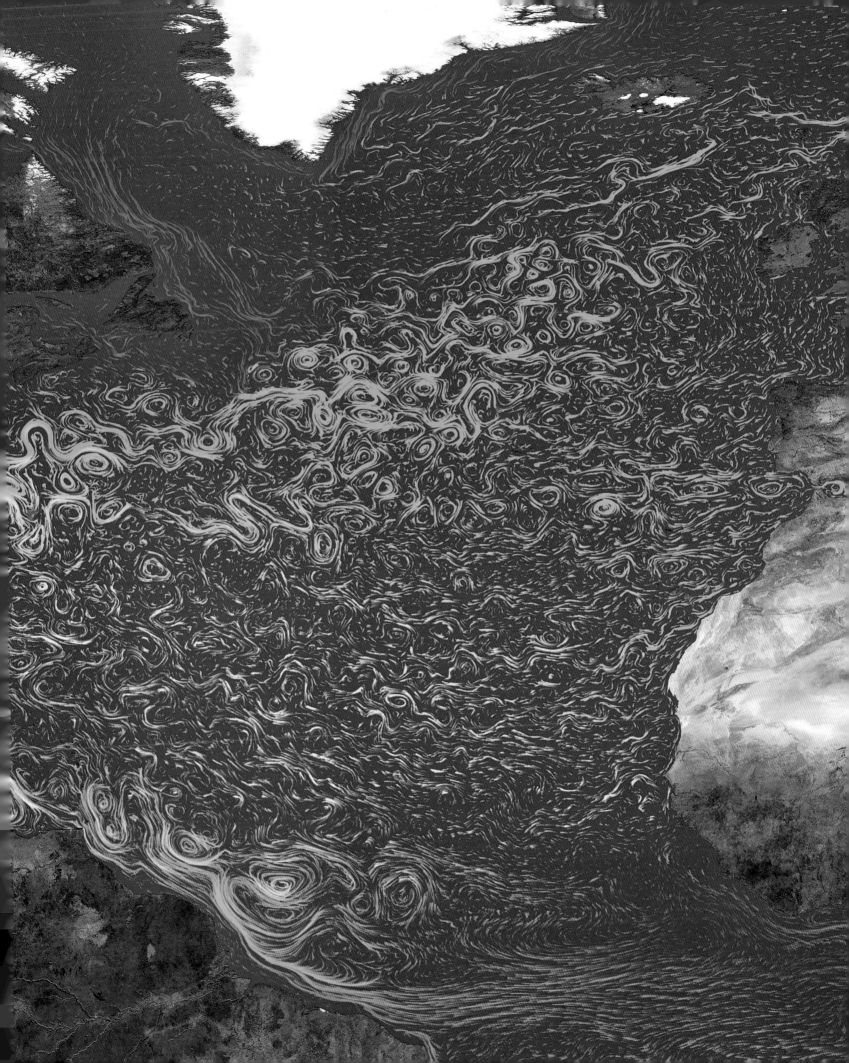

VERTICAL DIFFUSION

The slow upward diffusion of water from several hundred metres below the surface is called upwelling. The water rises to fill the space left by the offshore deflection of surface water, which is caused by a combination of wind direction, prevailing currents, and the Coriolis effect of Earth's rotation. Where winds and currents are reversed, the movement of water is towards the shore and the surface water is forced downwards. This is known as downwelling.

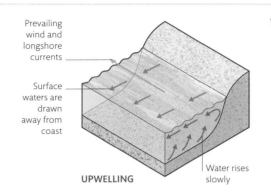

Prevailing wind and longshore currents

Surface waters are drawn away from coast

Water rises slowly

UPWELLING

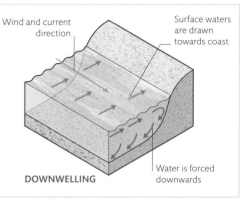

Wind and current direction

Surface waters are drawn towards coast

Water is forced downwards

DOWNWELLING

upwelling and plankton blooms

Upwellings are rising currents that bring water rich in chemical nutrients up from the depths. At the surface, minute photosynthetic organisms called phytoplankton build their bodies from these nutrients, their growth fuelled by solar energy. Phytoplankton, which include algae and cyanobacteria, are consumed by drifting animals known as zooplankton. Together, the two types of plankton support myriad marine grazers and predators.

Bloom of phytoplankton
In spring and summer, upwellings can trigger population explosions, or blooms, of phytoplankton. In this satellite view, a vibrant green phytoplankton bloom in the Baltic Sea is caught in swirling surface currents off Estonia. The central spiral eddy is about 30 km (20 miles) across. Such blooms can span hundreds or even thousands of kilometres.

Phytoplankton

Living out short lives in sunlit surface waters, phytoplankton are single-celled floaters and drifters. The majority, including most dinoflagellates, are microscopic; others, such as diatoms, are the size of sand grains. Many phytoplankton are master sculptors, building intricate skeletons out of calcium or silicon that they obtain from sea water.

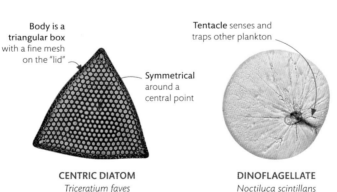

Body is a triangular box with a fine mesh on the "lid"

Symmetrical around a central point

CENTRIC DIATOM
Triceratium faves

Tentacle senses and traps other plankton

DINOFLAGELLATE
Noctiluca scintillans

Elongated body divided along its length into two symmetrical halves

PENNATE DIATOM
Pleurosigma angulatum

Zooplankton

Wherever phytoplankton bloom, zooplankton abound too, including some of the smallest of all marine animals. Zooplankton ride on tides and currents, although some are also able to swim very weakly. Holoplankton spend their entire lives as surface plankton; meroplankton are the larvae of organisms that complete their life cycles in deeper water.

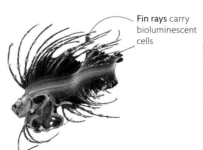

Fin rays carry bioluminescent cells

CUSK-EEL LARVA
Brotulotaenia nielseni

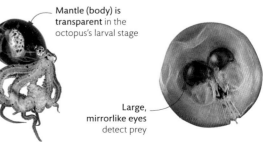

Mantle (body) is transparent in the octopus's larval stage

OCTOPUS LARVA
Wunderpus photogenicus

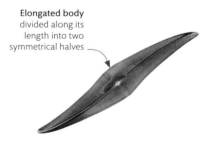

Large, mirrorlike eyes detect prey

OSTRACOD
Gigantocypris muelleri

deep currents

Incised into the steep continental slopes that rim the ocean basins (see p.198) are narrow gullies, deep gorges, and tightly meandering channels. All but those in shallow water (main picture) are completely hidden from view. Some are deeper than Arizona's Grand Canyon, others snake across the seabed for more than 3,000 km (1,860 miles). These channels funnel sediment from land to the deep ocean in underwater flows, known as turbidity currents, up to half-a-kilometre thick and travelling at speeds over 95 km/h (60 mph). Elsewhere, the ocean bottom is pockmarked with craters where methane gas has belched upwards from buried organic-rich sediment, and scoured by huge submarine slides whose displacement may have triggered powerful tsunamis in the water above. Deep currents sculpt the seabed into ripples, dunes, and giant sediment waves, tracing the path of the thermohaline conveyer belt (see panel, opposite) across the marine underworld.

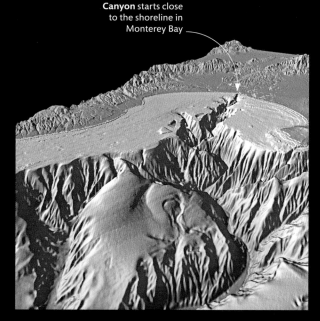

Canyon starts close to the shoreline in Monterey Bay

Seafloor relief, Monterey Bay, US
Monterey Canyon cuts a deep chasm across the Californian shelf before plunging down the steeply gullied slope. It extends for 470 km (290 miles) to the abyssal plain at a depth in excess of 4,000 m (13,000 ft).

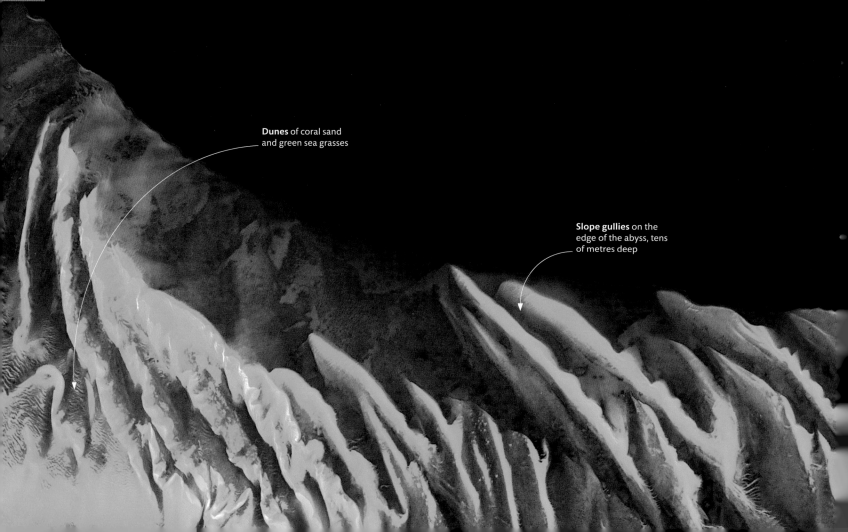

Dunes of coral sand and green sea grasses

Slope gullies on the edge of the abyss, tens of metres deep

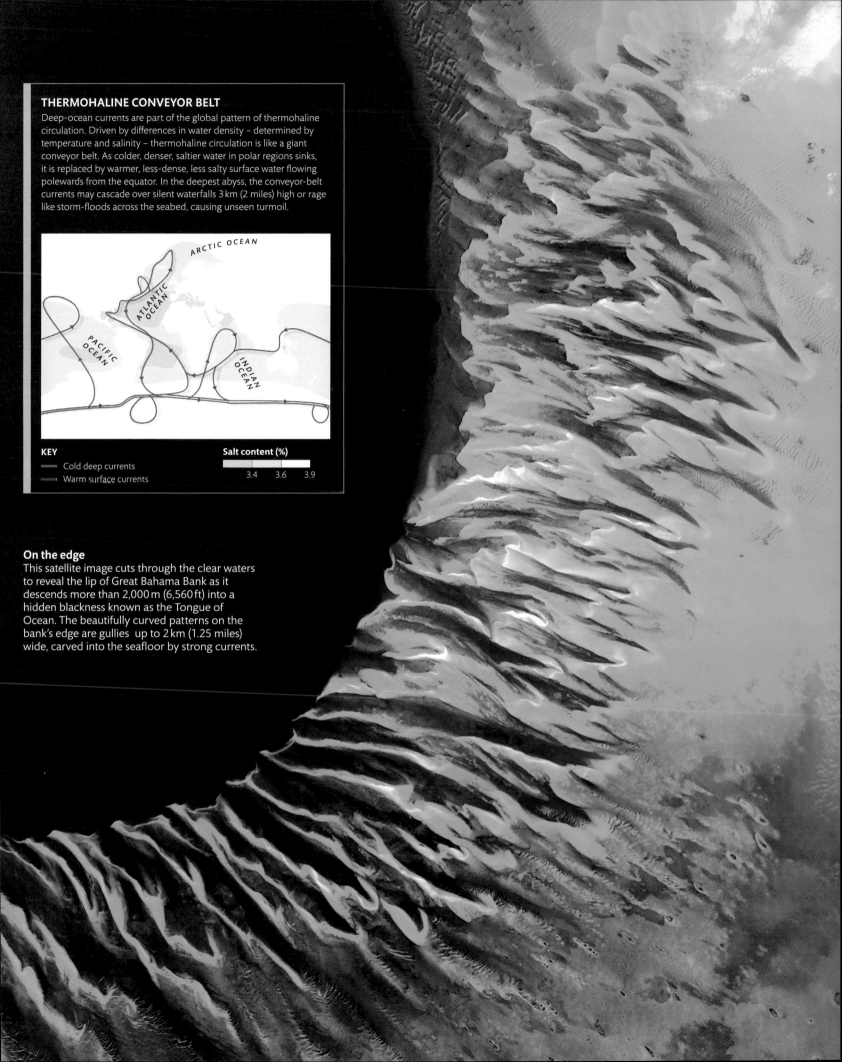

THERMOHALINE CONVEYOR BELT

Deep-ocean currents are part of the global pattern of thermohaline circulation. Driven by differences in water density – determined by temperature and salinity – thermohaline circulation is like a giant conveyor belt. As colder, denser, saltier water in polar regions sinks, it is replaced by warmer, less-dense, less salty surface water flowing polewards from the equator. In the deepest abyss, the conveyor-belt currents may cascade over silent waterfalls 3 km (2 miles) high or rage like storm-floods across the seabed, causing unseen turmoil.

ARCTIC OCEAN

ATLANTIC OCEAN

PACIFIC OCEAN

INDIAN OCEAN

KEY

Cold deep currents

Warm surface currents

Salt content (%)

3.4　3.6　3.9

On the edge

This satellite image cuts through the clear waters to reveal the lip of Great Bahama Bank as it descends more than 2,000 m (6,560 ft) into a hidden blackness known as the Tongue of Ocean. The beautifully curved patterns on the bank's edge are gullies up to 2 km (1.25 miles) wide, carved into the seafloor by strong currents.

Carboniferous sandstone stacks, about 340 million years old

Surfing the tidal bore, Alaska
A tidal bore occurs when a strong tide surges up an inlet or river, travelling against the current as a wave or series of waves. At Turnagain Arm, Alaska, bores can be 3 m (10 ft) high and move at 25 km/h (15 mph).

Flowerpot Rocks, Bay of Fundy
The greatest tidal range – over 16 m (50 ft) – occurs in Canada's Bay of Fundy. The sea covers and erodes the base of these sandstone stacks twice each day.

tides

The rise and fall of tides has occurred since the oceans formed around 4 billion years ago. Tides are waves with very long wavelengths that sweep around the planet. High tides are the crest of the wave and low tides the trough. Yet the reality of tidal motion is complex and variable due to frictional and topographic restrictions, the shape of ocean basins and marginal seas, and in recent times, human changes such as the dredging of rivers. The difference between high and low tides is the tidal range, which varies from zero to as much as 12–16 m (40–50 ft) in confined inlets and estuaries.

MONTHLY TIDAL CYCLE

The monthly tidal cycle is affected by the gravitational forces of the Earth–Moon and Earth–Sun systems. When Sun and Moon are aligned (at full moon and new moon), the two force fields act in the same direction, combining to yield very high and low spring tides. When they are at right angles (at first and last quarter), they pull in different directions, giving less extreme neap tides.

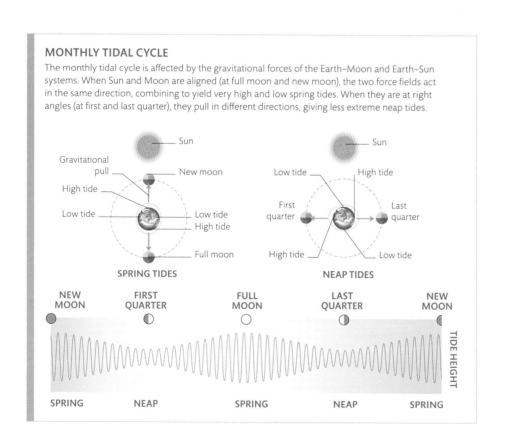

WAVE FORMATION

As winds blow across an area of sea, known as the fetch, waves develop. Small ripples, known as capillary waves, gradually build into chaotic choppiness. Beyond the fetch, wave interaction develops a distinctive regular pattern, called swell. As a wave enters shallow water, it slows against the seafloor. Its height increases until it spills forwards as a breaker.

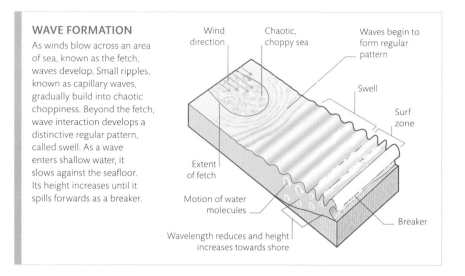

Wind direction

Chaotic, choppy sea

Waves begin to form regular pattern

Swell

Surf zone

Extent of fetch

Motion of water molecules

Breaker

Wavelength reduces and height increases towards shore

oceanic waves

As familiar and changeable as the colour of the sea, waves pound the shore and erode the land with their relentless power. Almost all waves are caused by wind stress acting at the sea's surface. The greater the wind force, the bigger the wave, and the more energy is stored and then released when the wave breaks. Individual storm waves produce instantaneous pressures of up to 30 tonnes per sq m (2.7 tons per sq ft), enough to destroy cliffs, piers, and coastal buildings. Storms at sea generate a train – a distinctive series of waves of a similar wavelength that, if not disrupted, retains its unique signature as it travels thousands of kilometres across an entire ocean.

Ship steers to face wave head-on; a full cargo keeps it steady as it rides the rogue wave

Rogue wave can reach a height of more than 15 m (50 ft)

Wave power

A rebounding wave spectacularly clashes with an incoming breaker, whipping up the surf at Cape Disappointment in Washington State, US. Wave energy and erosive power become focused on coastal headlands such as this, when the wave train encounters land (see p.195).

Rogue waves

Conditions at sea sometimes cause waves to merge and create a single, unusually large wave called a rogue wave. Rogue waves in the open ocean are isolated and unpredictable – and a danger even to large ships.

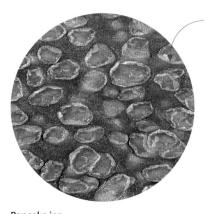

Elevated rims due to repeated collisions between ice pancakes

Ice floes
This jigsaw-like network of large ice floes, off Höfn in southeast Iceland, provides a temporary resting place for a seal, as well as a vantage point for hunting. Ice floes, which range in thickness from 0.5–5 m (1½–15 ft), repeatedly fuse into larger areas of sea-ice, and then break up again due to rough seas and partial melting.

Pancake ice
Ice discs 0.3–3 m (1–10 ft) in diameter with upturned edges are called pancake ice. They form under conditions of moderate to high wave activity.

frozen sea

Icebound, ice-cold, and inhospitable – such is the nature of polar seas. The North Pole sports an ice-covered ocean surrounded by land, the South Pole an ice-capped continent set amid frozen sea. In the dark polar winters, when temperatures drop below -30 °C (-22 °F), the sea freezes. Being less dense than sea water, the ice that forms floats. Around 7 per cent of the ocean's surface – an area roughly equivalent to the size of North America – is covered by floating sea-ice at any given time. Fast ice is sea-ice that is anchored to the shore, while pack ice drifts on currents. With the continuous daylight of polar summers, much of the ice melts back into the ocean. Some ice remains year-round, but global warming is reducing the area of permanent sea-ice cover.

HOW SEA-ICE FORMS

Freezing conditions cause ice crystals to develop near the sea's surface. As the wind blows, the crystals fuse into a slushy layer – grease ice – that hardens and thickens as temperatures fall. Rough seas break this layer into rounded shapes known as pancake ice, which then coalesce into masses called ice floes. Year on year, thicker sea-ice builds up, especially near the shore.

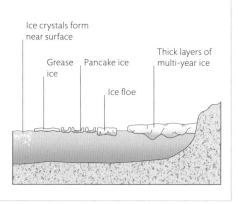

Ice crystals form near surface

Grease ice

Pancake ice

Thick layers of multi-year ice

Ice floe

Deceptive appearance
The ice raft on which these walruses (*Odobenus rosmarus*) are resting is the above-surface portion of an iceberg in Hudson Bay, Canada. Most of an iceberg, around 90 per cent, lies hidden below the waterline.

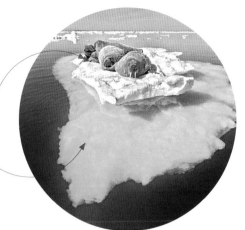

Walruses are insulated from the cold by blubber under the skin

Iceberg floats side-on, with its long axis parallel to the surface; it is only slightly less dense than the ocean, so it sits low in the water.

ice shelves and icebergs

Where flowing ice sheets and glaciers reach the coast they may extend over the sea as vast shelves of floating ice that tower above the waves. The ice in the shelves has accumulated from repeated snowfall over thousands of years and become progressively more compressed. As tiny air bubbles are squeezed out, the ice turns from white to a bluish hue. Marine algae frozen into the ice yield a vivid green, while rock flour – particles ground from rocks as the ice moved over land – produces reds, yellows, and browns. An iceberg is part of an ice shelf or glacier that breaks off and floats out to sea. Icebergs range from about 15 m (50 ft) in length to the size of a small country, such as Luxembourg.

Iceberg with chinstrap penguins
After a long winter, penguins such as chinstraps (*Pygoscelis antarcticus*) visit icebergs and await the food bounty triggered by the spring plankton bloom (see p.182). The melting icebergs themselves help to fuel plankton growth by releasing nutrients picked up from land.

HOW ICEBERGS FORM

Floating ice shelves and glaciers are buffeted by fierce winds and powerful waves, but even greater strain is caused by tides rising and falling beneath the ice. Cracks form in the ice and open up, while existing crevices are further enlarged. Eventually, in a process known as calving, large chunks break off and crash into the sea, floating away as blocks of freshwater ice called icebergs.

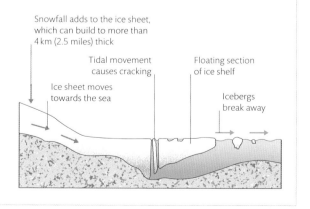

Snowfall adds to the ice sheet, which can build to more than 4 km (2.5 miles) thick

Tidal movement causes cracking

Floating section of ice shelf

Ice sheet moves towards the sea

Icebergs break away

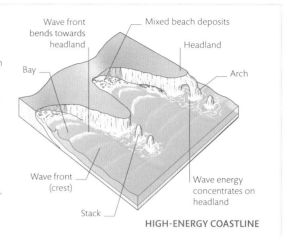

COASTAL EROSION

A typical high-energy coastline displays rapidly eroding cliffs, boulders, and headlands. When a wave front reaches the shore it refracts (bends) towards the shallower water adjacent to the headlands, where the energy becomes focused. Continuous erosion creates a negative sediment budget, with more material removed than added, and it carves out caves, arches, and stacks. Between headlands, beaches build up in sheltered bays, where more sediment is deposited than eroded.

Wave front bends towards headland

Mixed beach deposits

Headland

Bay

Arch

Wave front (crest)

Wave energy concentrates on headland

Stack

HIGH-ENERGY COASTLINE

where land meets the sea

A dynamic duel between land and sea is played out along the coast. Ocean water is the master sculptor and destroyer: it erodes cliffs, moulds the rocky shore, rounds pebbles, and grinds sand to create long, golden beaches that are eventually washed out to sea. But the coastline is maintained by a sediment supply from rivers of around 20 billion tonnes per year, and more is added from windblown dust and glacial discharge. Coastlines are as varied as they are stunning, from intricate fiords and sea-pounded towering cliffs to vast muddy deltas, mangrove swamps, and white coral sands.

Skeleton Coast, Namibia
Four of the world's longest beaches, each stretching over 100 km (60 miles), lie along the Skeleton Coast, where the Namib Desert meets the South Atlantic Ocean. The desert is around 60 million years old, and its dunes – some of which are 300 m (1,000 ft) tall – extend right up to the shore.

Ganges River Delta
This vast delta is part wild mangroves and part highly cultivated. Monsoon floods sweep its silty sediments out to sea, building a giant submarine delta 2,500 km (1,550 miles) long across the Bay of Bengal.

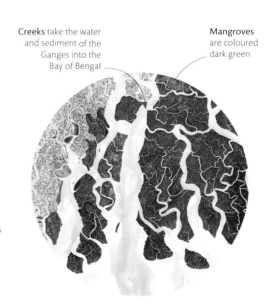

Creeks take the water and sediment of the Ganges into the Bay of Bengal

Mangroves are coloured dark green

Shingle beach is
30 m (100 ft) high

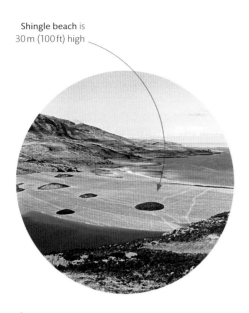

Uplifted beachfronts, New Zealand
In earthquake-prone regions, Earth's crust can be pushed either up or down by tectonic forces. This view (right) of Turakirae Head, New Zealand, shows a series of three ancient beach ridges snaking across the land above the present-day beach. The beaches were left behind between 160 years (the lowest) and 5,000 years ago (the highest). These represent the position of previous storm beachfronts that have been uplifted by major earthquake activity.

Rebounding land
Icecaps depress the underlying landmass, which slowly rebounds when the ice melts. Such a process, called isostatic rebound, created this raised shingle beach (left) at Loch Tarbert on the Isle of Jura, Scotland.

sea-level change

Sea level is the mean height of the ocean surface relative to land. This has varied throughout Earth's history in response to tectonic movements and climate change. When global temperatures were much colder, more water was locked up in icecaps and glaciers, causing sea level to drop by up to 120 m (400 ft) four times in the past million years. Broad continental shelves became dry land. Longer ago, the rate of ocean spreading and mountain uplift caused even greater change – sea level was 250–350 m (820–1,150 ft) higher and ocean covered 82 per cent of the world. Global warming today is expanding sea water and melting icecaps. Sea level may rise 30–50 cm (12–20 in) by the century's end.

EVIDENCE FOR SEA-LEVEL CHANGE
This frontispiece of the ninth edition of Charles Lyell's 1853 *Principles of Geology* shows the Roman Temple of Serapis (now re-interpreted as a market place) in Pozzouli, southern Italy. Lyell argued that a distinctive layer of boreholes made by marine *Lithophaga* bivalves in the temple columns indicated that they were once submerged under the sea and then later uplifted, thus supporting the idea of sea-level change and the Uniformitarian theory of geology. Recent research broadly validates his conclusions.

Band of boreholes left in the marble columns by marine bivalves

TEMPLE OF SERAPIS, POZZOULI, ITALY

SHALLOW SEA TO DEEP OCEAN

Continents are surrounded by gently sloping continental shelves. Beyond the shelves lie steeper slopes marked by deep canyons and drainage channels. Vast underwater rivers called turbidity currents and massive slides, or slumps, carry sediment down the slope. Deposited on the continental rise, sediment is reworked into sideways drifts by powerful contour currents, or spills across the deep-ocean basin as submarine fans.

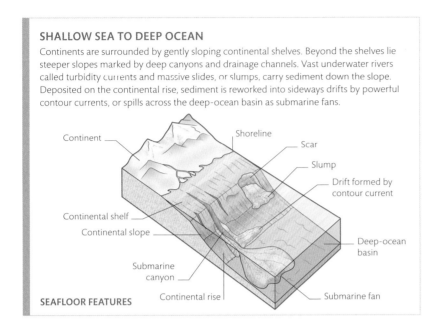

Continent
Shoreline
Scar
Slump
Drift formed by contour current
Continental shelf
Continental slope
Deep-ocean basin
Submarine canyon
Continental rise
Submarine fan

SEAFLOOR FEATURES

shallow seas

Shallow seas extend from the coast to the edge of the continental shelf, at about 100 m (300 ft) water depth. Bathed in sunlight and often rich in nutrients, they are ideal for photosynthesis and the growth of coral reefs and kelp forests. They also provide nursery habitats for juvenile organisms and feeding grounds for a host of marine species. Shallow seas are shaped both by their biogenic activity and by the dynamic effect of waves, tides, ocean currents, and coastal upwelling. Ripples and dunes covering the seafloor are testament to the transport of sediment from land across the continental shelf.

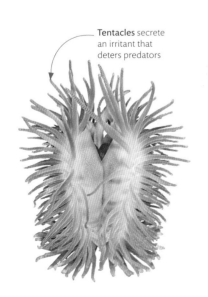

Tentacles secrete an irritant that deters predators

Seabed engineer
Flame clams (*Limaria hians*) effect major transformations of the seabed. The nests they build from shells and stones combine to form a dense, reeflike structure that raises and stabilizes the seabed, creating a habitat that supports a rich community of organisms.

Kelp forest, offshore California
Forests of giant kelp (*Macrocystis pyrifera*) grow in nutrient-rich, temperate shelf waters, such as those off California, US. Anchored by holdfasts and kept afloat by gas bladders, the kelp can grow up to 60 cm (2 ft) per day. Many ocean animals find safe haven beneath the kelp forest's canopy.

World ocean floor
Marie Tharp published this map of the
world ocean floor in 1977, shortly after
her colleague, Bruce Heezen, died at
sea while on a new oceanographic
expedition in the Atlantic Ocean.

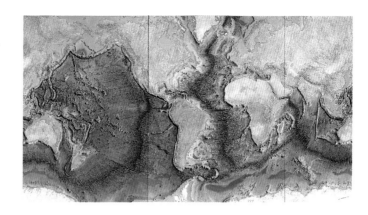

history of Earth science

seafloor mapping

Marie Tharp at work
Pictured here at the Lamont Geological Observatory,
New York, US, Marie Tharp painstakingly transcribed
sonar readings to create her ground-breaking map of
the world ocean floor.

Probably the most important map of the 20th century was a completely new view of the seafloor, painstakingly compiled from 1957 until 1977 by the American geologists Marie Tharp and Bruce Heezen. For the first time, the ocean floor – which comprises 71 per cent of Earth, and which had previously been largely unknown – was revealed to the public and scientists alike, leading to fundamental insights into the nature of our planet.

The Heezen-Tharp maps captured on paper the topography of the seafloor. The maps revealed the greatest mountain range on Earth – the mid-ocean ridge system – which rises 3,000 m (10,000 ft) above the seafloor and exends 60,000 km (37,000 miles) around the planet. They showed dark chasms plunging to a depth of over 1 km (0.6 miles). Deep canyons fringing the landmasses incise the flat shelves, feeding sediments eroded from land to gigantic submarine fans sprawled across flat abyssal plains. The maps revealed countless submerged seamounts as well as active volcanoes rising to the surface, constructing reef-fringed islands. Their work provided key evidence for plate tectonic theory, which was not widely accepted until the 1960s.

Tharp and Heezen used sonar readings of the seafloor to create their maps. Their map of the world ocean floor and its beautiful 1977 rendition painted by Heinrich Berann was a true landmark.

Today, our methods are ever more sophisticated and precise, using satellite imagery to refine the global ocean map, and high-precision deep-towed sonar to resolve a myriad seafloor features. These include sediment waves, gas-escape pockmarks, deepwater coral reefs, and even the fault-displacement in the Sumatran trench, which was the cause of the devastating 2004 tsunami.

The map that shook the world
Tharp and Heezen's map, which revealed the Atlantic Mid-Ocean Ridge (opposite), revolutionized ocean science, but initially met with scepticism. Jacques Cousteau's subsequent exploration of the ridge in a submersible, intended to disprove their findings, confirmed the existence of the rift valley where seafloor spreading occurred.

> " It became clear that existing explanations for the formation of the Earth's surface no longer held. "
>
> HALI FELT, *SOUNDINGS: THE STORY OF THE REMARKABLE WOMAN WHO MAPPED THE OCEAN*, 2012

Hunters of the open ocean

Albatrosses are pelagic predators that spend many months, sometimes more than a year, on the open sea, before returning to remote islands to breed. These black-browed albatrosses (*Thassalarche melanophrys*) scour the turbulent, nutrient-rich waters of the Southern Ocean for prey.

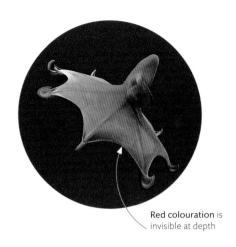

Red colouration is invisible at depth

Denizen of the deep

The deep ocean is a dark, cold, high-pressure habitat. Dumbo octopuses (*Grimpoteuthis* spp.) live at depths of up to 4,000 m (13,000 ft) or more. They hover over the seafloor in search of bivalves, crustaceans, and worms.

open ocean waters

In the open ocean – a vast expanse covering two-thirds of the planet – primary productivity is low and the waters are sparsely inhabited. The ocean is stratified into layers of different temperature, salinity, and light penetration, and compartmentalized by currents. When such subtle barriers are disturbed by storm stirring, local upwelling, and the meeting of water masses, dissolved nutrients rise to the surface and promote phytoplankton blooms. Grazers gather to feed, but there is nowhere to hide. Many strategies have evolved to avoid predation: transparent bodies and countershading, mass migrations of zooplankton from the surface to deep water, and swimming in large schools.

OCEAN THERMOCLINE

The thermocline marks the base of a warm surface layer, 50–1,000 m (150–3,300 ft) thick, throughout temperate and tropical oceans. In the tropics, calm weather stabilizes the thermocline, inhibiting mixing so that surface waters are gradually depleted of nutrients. Storms in temperate mid-latitudes enhance mixing. This stirs the thermocline and brings nutrients to the surface: the green, phytoplankton-rich waters proliferate with life.

Warmer surface waters

Intense sunshine

Strong thermocline

Colder, deeper waters (no upward mixing)

Some phytoplankton sink to the bottom

TROPICAL OCEAN

Storm disturbs thermocline

Strong winds

Intense mixing

Abundant phytoplankton

TEMPERATE OCEAN

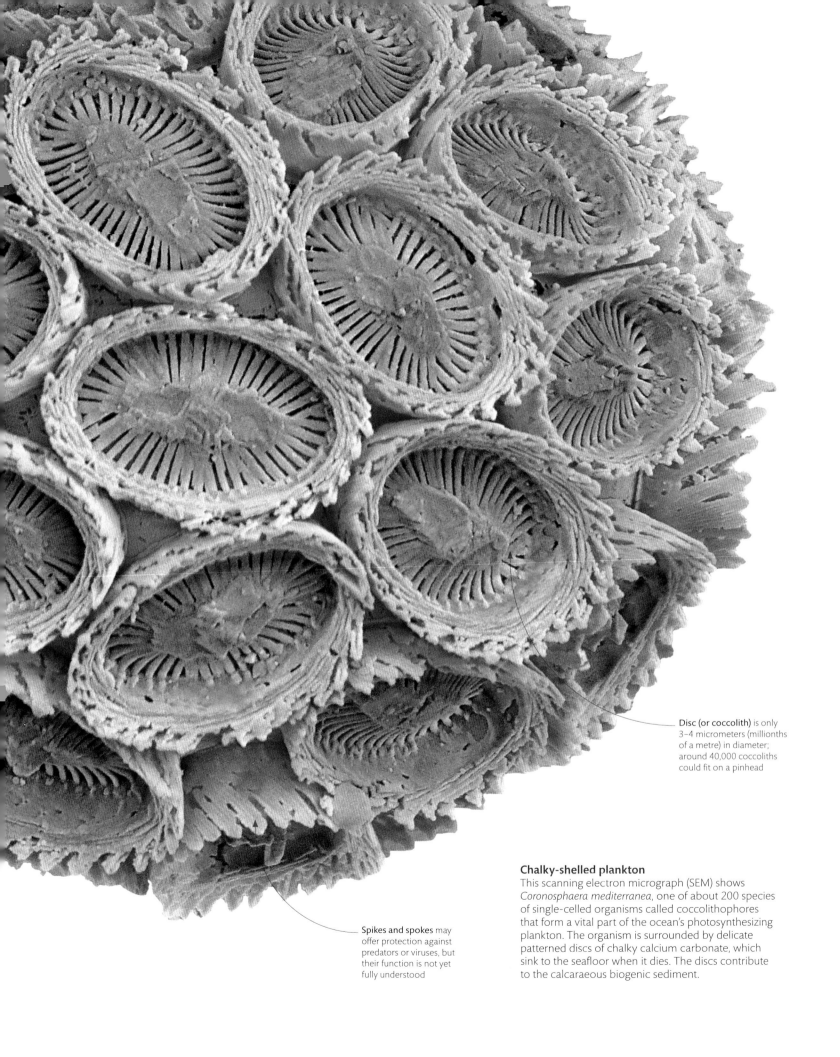

Disc (or coccolith) is only
3–4 micrometers (millionths
of a metre) in diameter;
around 40,000 coccoliths
could fit on a pinhead

Spikes and spokes may
offer protection against
predators or viruses, but
their function is not yet
fully understood

Chalky-shelled plankton

This scanning electron micrograph (SEM) shows
Coronosphaera mediterranea, one of about 200 species
of single-celled organisms called coccolithophores
that form a vital part of the ocean's photosynthesizing
plankton. The organism is surrounded by delicate
patterned discs of chalky calcium carbonate, which
sink to the seafloor when it dies. The discs contribute
to the calcaraeous biogenic sediment.

patterns on the seafloor

The seafloor is almost entirely covered with sediment. On the flanks of mid-ocean ridges, the sediment is a thin veneer over oceanic crust, but along continental margins and beneath major deltas, its thickness can exceed 16 km (10 miles). Each sediment particle tells a different tale of process, climate, and environment. Some sediment grains are carried from land to seafloor by wind, rivers, glaciers, and underwater currents. Others, biogenic particles, are the remains of planktonic organisms such as foraminifera, coccolithophores, radiolarians, and diatoms (see p.182). Sediments can also contain mineral and metal deposits that precipitate out of concentrated brines or from volcanic emissions, and even space-dust that falls to the ocean floor.

Foraminiferan microfossils
The size of medium to coarse sand grains, foraminifera are single-celled protozoans, most with tests (shells) of calcium carbonate. Of more than 50,000 known species, the majority live on the seafloor; others are key members of the zooplankton.

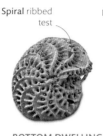

Spiral ribbed test

BOTTOM DWELLING
(*Elphidium crispum*)

Pores allow armlike pseudopodia to reach out and catch prey

PLANKTONIC
(*Orbulina universa*)

Test is divided into several chambers

BOTTOM DWELLING
(*Cribroelphidium poeyanum*)

Radiolarian microfossils
Like foraminifera, radiolarians are zooplankton, or animal-like plankton. These tiny protozoan predators of the open ocean range from 0.03 to 2 mm ($^{1}/_{12}$ in) across. They have a glassy silica test with pores through which they project fingers of cellular material called pseudopodia.

Many regular pores

SPHERICAL OPEN FORM
(*Theocapsa* spp.)

Varied pore size

SMOOTH NUT-SHAPED FORM
(*Tripospyris* spp.)

Sharp spines on rim

HELMET-SHAPED FORM
(*Anthocyrtidium ligularia*)

SEAFLOOR SEDIMENTS
Around land and near the poles, the seabed is blanketed by sediment from glaciers and other agents of land erosion. Siliceous sediments (made of silica) occur in belts where upwelling encourages blooms of silica-rich plankton. Calcareous sediments (made of calcium carbonate) also reflect the rain of microscopic tests from the plankton, but they are preserved only above a depth of 4–5 km (2.5–3 miles). Below this level, more acidic waters dissolve the calcium carbonate, leaving just fine material called abyssal red clay.

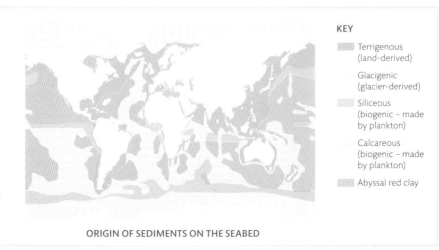

KEY

- Terrigenous (land-derived)
- Glacigenic (glacier-derived)
- Siliceous (biogenic – made by plankton)
- Calcareous (biogenic – made by plankton)
- Abyssal red clay

ORIGIN OF SEDIMENTS ON THE SEABED

HYDROTHERMAL VENTS

Cold sea water percolating into cracks and fissures on newly formed seafloor penetrates deep into the oceanic crust, leaching out a cocktail of chemicals as it seeps downwards. When the water encounters hot magma, it is superheated and forced back to the surface, emerging from the vent as a hot spring. Dissolved minerals precipitate as black smokers, depositing metal-rich sediments and constructing a "cityscape" of chimneys, spires, and stacks. Communities of clams, mussels, barnacles, anemones, limpets, crabs, worms, shrimps, and fish surround the vents, entirely supported by microbes.

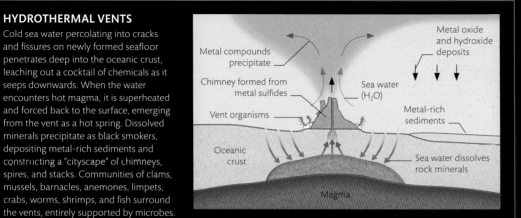

Metal compounds precipitate

Chimney formed from metal sulfides

Vent organisms

Metal oxide and hydroxide deposits

Sea water (H_2O)

Metal-rich sediments

Oceanic crust

Sea water dissolves rock minerals

Magma

MINERAL DEPOSITION AT A HYDROTHERMAL VENT

islands in the ocean

Islands are like marine oases – havens for terrestrial and aquatic life. They can also tell us about oceans past and present. The hundreds of thousands of islands in the world's oceans can be classified into four main types. Volcanic islands and coral reefs are often smaller and ephemeral, subject to erosion, sinking, and biotic changes. Continental islands, which vary in size, are fragments splintered off the main landmass by a rise in sea level or by tectonic separation. Complex islands are larger and long-lived, formed by a combination of processes; some, such as Cyprus in the Mediterranean, have a core of deep-ocean crust, known as an ophiolite, which has been squeezed to the surface between colliding tectonic plates.

Sea water discoloured by volcanic ash and algae

New volcanic island

Nishinoshima

Merging islands
Japan's volcanic Nishinoshima Island grew by merging with a new island that arose after eruptions in 2013. Nishinoshima lies over a subduction zone (see pp.114–15) where the Pacific plate slides under the Philippine plate.

EVOLUTION OF VOLCANIC ISLANDS

Some volcanic islands, such as the Hawaiian and Galápagos Islands, emerge above hotspots from mantle plumes (see p.138); others, including those of Tonga and the Mariana Islands, form over subduction zones (see pp.114–15). Born from extensive outpourings of lava that build up on the seafloor, a volcanic island becomes explosive as it breaks the surface. If it survives early erosion, it may grow into a substantial cone, fringed in tropical waters by coral reefs. When the lava flow stops, the island subsides to form a seamount. As the island sinks, the corals continue growing upwards towards the sunlit shallows, producing a ring-shaped reef at the surface called an atoll.

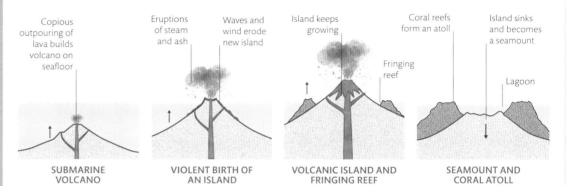

| Copious outpouring of lava builds volcano on seafloor | Eruptions of steam and ash | Waves and wind erode new island | Island keeps growing | Coral reefs form an atoll | Fringing reef | Island sinks and becomes a seamount | Lagoon |

SUBMARINE VOLCANO

VIOLENT BIRTH OF AN ISLAND

VOLCANIC ISLAND AND FRINGING REEF

SEAMOUNT AND CORAL ATOLL

Coral atoll, Indonesia

Pulau Ndaa island, in the Banda Sea, is an atoll – a coral reef surrounding a lagoon. The atoll sits atop an inactive volcano that has subsided and become a seamount. If an atoll's coral reef grows fast enough to keep pace with the volcano's subsidence, the atoll remains at the surface; if not, it sinks with the volcano, disappearing beneath the waves.

LAYERS OF THE ATMOSPHERE

The atmosphere has distinct layers with their own characteristics and phenomena. The troposphere, the lowest layer, is the location of much of Earth's weather systems and commercial aircraft; the concentration of ozone gases peaks within the stratosphere; and the mesosphere and thermosphere play host to dazzling meteor showers and dramatic aurorae. The Kármán Line marks the legal edge of space – the official upper edge of a country's airspace. The thermosphere and exosphere beyond are too thin to support aircraft.

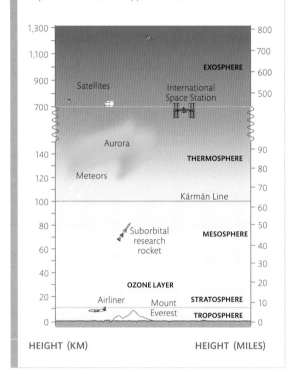

| HEIGHT (KM) | HEIGHT (MILES) |

The Leonids comprise an annual meteor shower made by debris left by the Tempel-Tuttle comet

METEOR SHOWERS

Meteoroids are small rocky or metallic objects in outer space. If these objects enter Earth's atmosphere, they burn up, producing streaks of light called meteors. An estimated 25 million meteoroids, space dust, and other debris enter Earth's atmosphere each day.

THE EDGE OF SPACE

In addition to its key component gases, the lowest layer of the atmosphere (the troposphere) contains water vapour and weather systems driven by solar heating. The atmosphere gradually thins with increasing distance from Earth, until it is indistinguishable from the interplanetary medium – outer space.

Earth's atmosphere

First formed from gases spewed out by the volcanoes of primitive Earth (see pp.28–29), our planet's ocean–atmosphere couplet is unique in the Solar System. Each of the atmosphere's layers comprises about 78 per cent nitrogen, 20.9 per cent oxygen, 0.9 per cent argon, and traces of 10–15 other gases. In the lower layers, the gas ozone absorbs ultraviolet radiation from the Sun, which would otherwise damage plant and animal DNA, and prevent plants from being able to carry out photosynthesis. Greenhouse gases, including carbon dioxide and methane, keep the planet at a habitable temperature by trapping infrared radiation. However, human activities have increased the amount of these gases in the atmosphere to dangerous levels, leading to global warming.

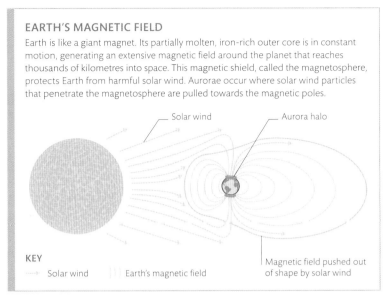

EARTH'S MAGNETIC FIELD

Earth is like a giant magnet. Its partially molten, iron-rich outer core is in constant motion, generating an extensive magnetic field around the planet that reaches thousands of kilometres into space. This magnetic shield, called the magnetosphere, protects Earth from harmful solar wind. Aurorae occur where solar wind particles that penetrate the magnetosphere are pulled towards the magnetic poles.

Solar wind

Aurora halo

KEY

→ Solar wind

)|| Earth's magnetic field

Magnetic field pushed out of shape by solar wind

aurorae

Earth's atmosphere is under constant bombardment from the solar wind – a flow of charged electrons and protons fired out from the Sun. Most of these particles are deflected by the planet's magnetic field, but some penetrate it near the magnetic poles. When these particles collide with nitrogen and oxygen atoms in the atmosphere, energy is released in the form of light, creating a dazzling display of dancing lights across the sky. These light shows are known as aurora borealis, or northern lights, in the northern hemisphere and aurora australis, or southern lights, in the southern hemisphere.

Northern lights
The colours emitted by an aurora depend on the type of atom struck by solar wind particles and the altitude at which this occurs. Green and red shades, pictured here over Iceland, are most common, but yellow, purple, blue, and pink hues are also possible.

Aurora australis seen from space encircling the South Pole

Polar halo
Most aurorae, like this one around Antarctica, occur in a band (halo) 3–6° wide located between 10–20° from the North or South Poles.

wind

From light breezes to hurricanes, air in the troposphere (see pp.210–11) is constantly moving around Earth due to differences in pressure and temperature. As the Sun heats different parts of the planet, the air around these warm spots heats up too. The resulting warm air rises into the sky, leaving an area of low pressure in its place. Cool air, which is more dense, sinks and fills the space. This constant circulation of air from regions of high pressure to regions of low pressure is what generates wind. These lower-atmosphere winds, called planetary winds or prevailing winds, form part of three great wind cells that dominate each hemisphere (see right). Local winds, which occur over a smaller area and time period, are the result of sea-breeze and land-breeze cycles in coastal areas.

Relentless winds
Prevailing westerly winds known as the Roaring Forties stunt and shape the growth of these trees at Slope Point on the southern tip of New Zealand. The powerful winds blow in the southern hemisphere between the latitudes of 40° and 50°, relatively uninterrupted by landmasses that could decrease their speed. The constant bombardment of air from one direction dries and kills the shoots and buds on one side of these trees, preventing new growth, while the other side is able to continue to grow as normal, creating the illusion of them bending in the breeze.

ATMOSPHERIC CIRCULATION

At the equator, intense solar heating warms the air, which then rises. More air is sucked in from either side of the equator to replace the rising air, causing winds, known as trade winds, to blow across the tropics. The rising air races poleward and cools rapidly, before sinking. This constant circulation generates a cell of air, called a Hadley cell, either side of the equator. Further pairs of cells occur at mid-latitudes and polar regions, causing prevailing westerlies and polar easterlies to blow.

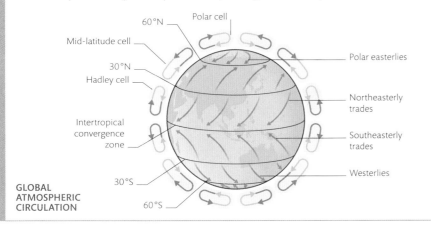

60°N
Polar cell
Mid-latitude cell
30°N
Hadley cell
Intertropical convergence zone
Polar easterlies
Northeasterly trades
Southeasterly trades
Westerlies
30°S
60°S

GLOBAL ATMOSPHERIC CIRCULATION

Dust walls can be up to 3 km (1.8 miles) tall.

Haboob

A haboob, like this one approaching Khartoum, Sudan, is an intense wind storm common in arid regions around the world. The strong wind can approach with little or no warning, whipping up a wall of sand and dust that can reach up to 100 km (60 miles) wide.

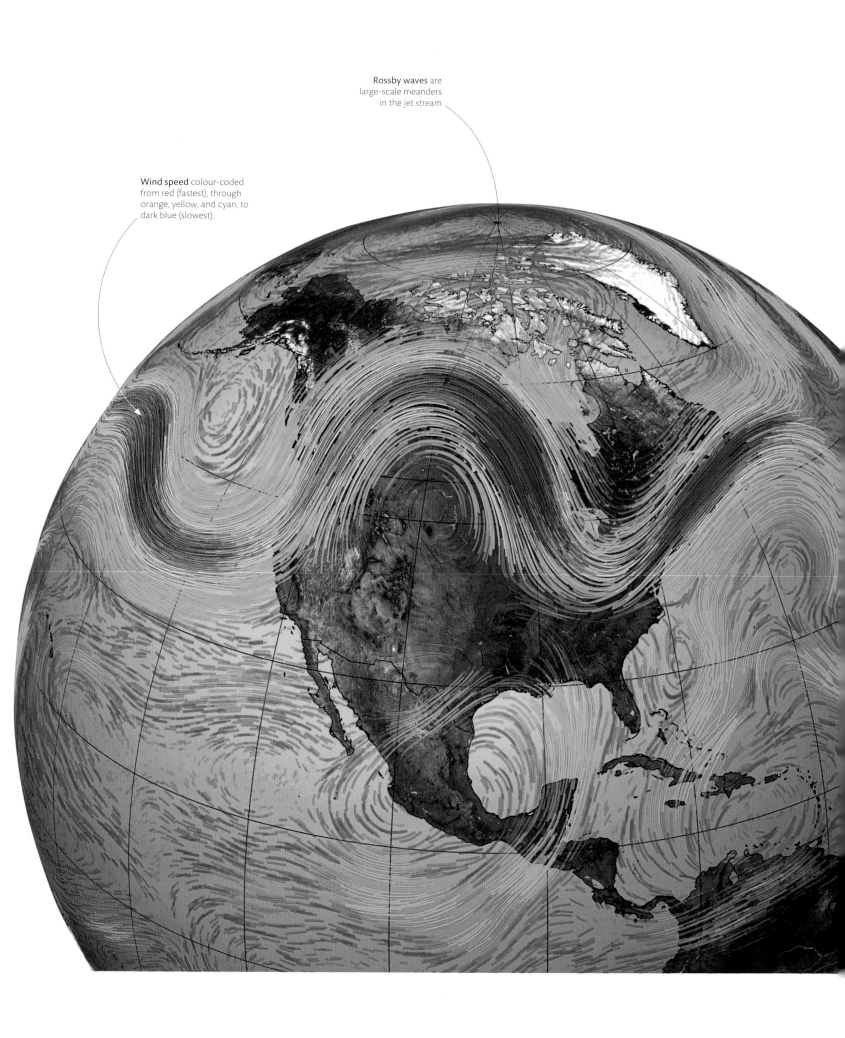

Rossby waves are
large-scale meanders
in the jet stream

Wind speed colour-coded
from red (fastest), through
orange, yellow, and cyan, to
dark blue (slowest).

The ozone hole
Coloured satellite maps (right) reveal depleted ozone levels in the atmosphere over Antarctica. Dark blue colouration shows areas of maximum depletion, known as the ozone hole. The hole continued to grow after 1987, reaching 28 million sq km (11 million sq miles) in 2000, but then stabilized and reduced.

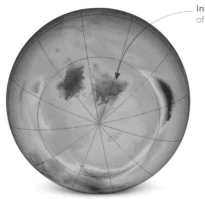

Initial discovery of two small holes

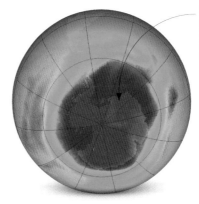

Enlarged ozone hole covering 26 million sq km (10 million sq miles)

1979

1987

history of Earth science

atmospheric imaging

Since the 1960s, the deployment of satellites carrying instruments to monitor Earth's atmosphere has allowed us to observe large parts of the world simultaneously, as well as to monitor specific parts over a longer time. Satellite imagery has made it possible to identify holes in the ozone layer, prompting action to address the problem, and has helped us to understand the atmospheric changes that are likely to affect life on Earth.

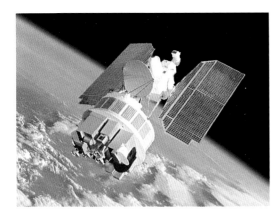

Eyes in space
Sensors aboard the historic satellite, Nimbus-7, detected abnormally low concentrations of ozone over the Antarctic in 1979–80, helping to identify the ozone hole.

Visualizing the jet stream
This modelling of the jet stream, the high-speed belt of eastward-moving winds in the upper troposphere, is based on satellite observations of water vapour levels over North America. The position of the jet stream plays a major role in determining the weather below, causing frequent storms and other weather phenomena associated with low pressure.

Satellites are routinely used to inform all aspects of ocean and atmospheric sciences. They have become essential for weather observation and prediction, and provide warning of extreme events. Long term observations chart the effects of atmospheric pollution from volcanoes, wild fires, and urban traffic, and help us to understand the nature and effects of natural climate oscillations, such as El Niño (see p.243).

To monitor weather, satellites are positioned in high geo-stationary orbits – nearly 36,000 km (22,400 miles). Most other scientific remote sensing is made from altitudes of 160–2,000 km (100–1,250 miles). These "low Earth orbits" have various inclinations to the equatorial plane, circling Earth every 90–120 minutes but covering a different track with each orbit. Complete coverage of the globe is possible in about 12 hours.

From the 1970s on, satellites have observed the atmosphere using instruments sensitive to infrared, the visible spectrum, and microwave bands. Some scan sideways to look at the upper atmosphere edge-on and can measure vertical variation in gases, temperature, and pressure. Others measure high-altitude winds by sensing Doppler shift.

To analyse atmospheric composition, mass spectrometry sensors identify gases by the way they absorb or emit particular radiation wavelengths. It was spectrometry data from satellites such as Nimbus-7 that was vital for identifying the thinning of the ozone layer. This led to the signing of the Montreal Protocol in 1987, which triggered action to stabilize and reduce the ozone hole. Another key contribution of satellite imaging is the measurement of levels of greenhouse gases to help us monitor the effect of efforts to combat climate change.

   Satellite images… have revolutionised our understanding of the ocean-atmosphere…

PROFESSOR IAN ROBINSON, NATIONAL OCEANOGRAPHY CENTRE, SOUTHAMPTON, UK

weather systems

Sunshine or cloud, rain or snow, strong winds or still calm – these changing local weather conditions are the result of solar energy interacting with the oceans, atmosphere, and land. Sunlight heats Earth more in some places than others, leading to differences in air pressure and the generation of wind (see pp.214–15). It also drives the water cycle (pp.224–25). Different parts of the globe experience a variety of weather systems – tropical, monsoonal, temperate, and polar. Temperate regions are subject to mid-latitude depressions (or cyclones) that build and dissipate over a period of days, travelling from west to east one after another. These are areas of low-pressure generated by Rossby waves – giant meanders in the high-altitude jet-stream winds, created by Earth's rotation (pp.216–17).

East coast bomb cyclone
This satellite image of the east coast of the US shows the dramatic spiralling clouds of a bomb cyclone. This is caused by a dramatic drop in air pressure from outside to inside a weather system, created by the collision of cold continental air and warm oceanic air masses.

Mid-latitude cyclone

Driven by prevailing westerly winds, fast-moving cold polar air dives beneath the slower moving warm tropical air, which is pushed upwards, producing an area of low pressure. In the northern hemisphere (pictured), air flowing into this area creates an anticlockwise spin. In the southern hemisphere, the spin is clockwise.

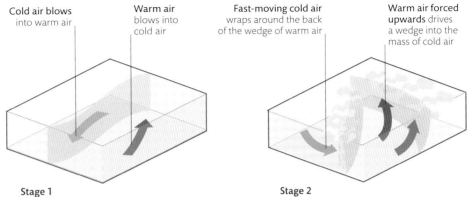

Cold air blows into warm air

Warm air blows into cold air

Fast-moving cold air wraps around the back of the wedge of warm air

Warm air forced upwards drives a wedge into the mass of cold air

Stage 1
As a cold air mass advances towards a warm one, the air masses start to interact. This is the initial stage of the process of cyclone formation.

Stage 2
Cold air pushes warm air upwards. The resulting rotation of both air masses starts to create a spiralling movement.

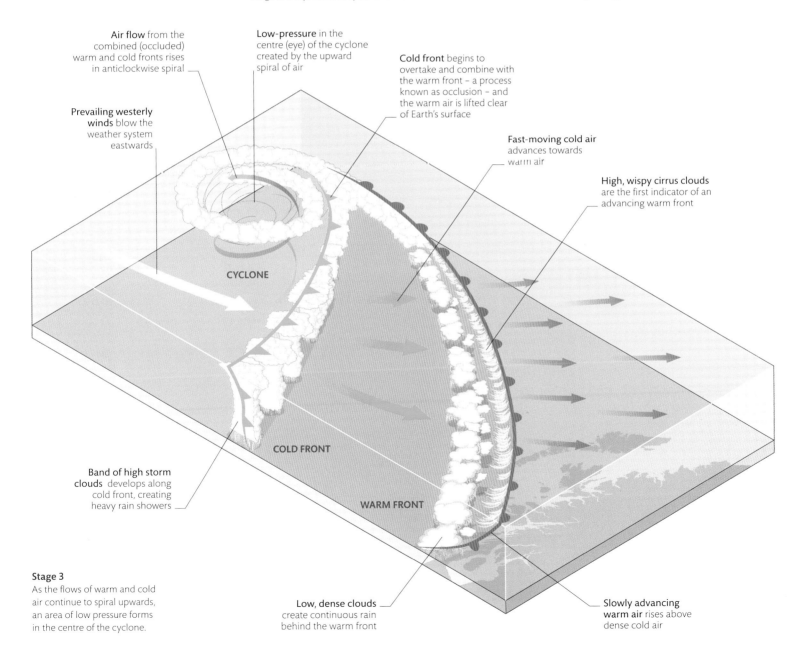

Air flow from the combined (occluded) warm and cold fronts rises in anticlockwise spiral

Low-pressure in the centre (eye) of the cyclone created by the upward spiral of air

Cold front begins to overtake and combine with the warm front – a process known as occlusion – and the warm air is lifted clear of Earth's surface

Fast-moving cold air advances towards warm air

High, wispy cirrus clouds are the first indicator of an advancing warm front

Prevailing westerly winds blow the weather system eastwards

CYCLONE

COLD FRONT

WARM FRONT

Band of high storm clouds develops along cold front, creating heavy rain showers

Stage 3
As the flows of warm and cold air continue to spiral upwards, an area of low pressure forms in the centre of the cyclone.

Low, dense clouds create continuous rain behind the warm front

Slowly advancing warm air rises above dense cold air

Eye of the storm

While there is a calm area of cool air that descends at the centre of a storm – called the eye, seen here in Hurricane Florence in 2018 – the strongest winds in a tropical cyclone spiral upwards through the eyewall. The highest thunderclouds, powerful storms, and frequent lightning strikes surround the eye.

The calm eye region is typically around 32–64 km (20–40 miles) across.

tropical cyclones

Tropical cyclones are the largest, most violent storms on Earth, with sustained wind speeds of 120–250 km/h (75–155 mph) and record gusts of over 400 km/h (250 mph). Those that originate in the north Atlantic and northeast Pacific oceans are called hurricanes; those that begin in the northwest Pacific are called typhoons; and those that form in the south Pacific and Indian oceans are known simply as cyclones. The energy released by a single tropical cyclone is equivalent to about 10,000 nuclear bombs. Although they form over warm tropical seas, when these violent, rotating storms reach coasts and travel inland they can inflict widespread devastation – their combination of high winds and excessive rainfall tear down homes and trees, and cause catastrophic storm surges.

TROPICAL CYCLONE FORMATION

Tropical cyclones develop where sea surface temperatures exceed 27°C (80°F). Water evaporates from the overheated surface, rises, and condenses to form cumulonimbus thunderclouds that billow upwards to the top of the troposphere (see pp.210–11), 12–16 km (6–10 miles) high.

Cool air rushes inwards to fill the void created by the rising, humid air, and Earth's rotation causes the storm winds to spin anticlockwise in the northern hemisphere and clockwise in the southern hemisphere. Cool, dry air sinks through the cloud-free eye, a region of calm at the centre of the storm.

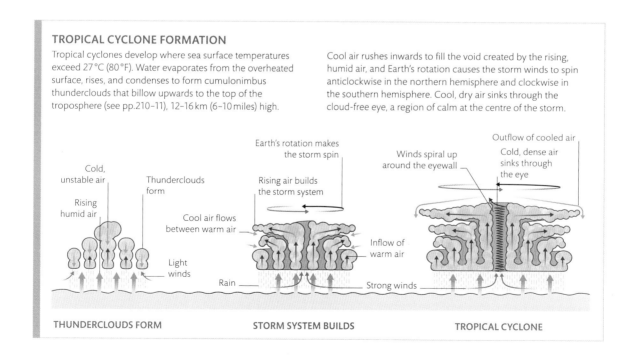

Cold, unstable air
Rising humid air
Thunderclouds form
Light winds

Cool air flows between warm air

Earth's rotation makes the storm spin
Rising air builds the storm system
Rain
Inflow of warm air

Winds spiral up around the eyewall
Outflow of cooled air
Cold, dense air sinks through the eye
Strong winds

THUNDERCLOUDS FORM   STORM SYSTEM BUILDS   TROPICAL CYCLONE

Atlantic hurricanes
Throughout the summer, sea surface temperatures build up in the tropical North Atlantic and Caribbean region, spawning a six-month hurricane season from June through November. Like many other hurricanes that hit the area, Hurricane Joaquin (seen here) caused widespread damage as it swept through the Caribbean in October 2015.

High-level to extreme-level clouds

While some types of cloud, including noctilucent and nacreous clouds, are found as high as the mesosphere and stratosphere (see p.210), most form in the troposphere. Within the troposphere, the highest-altitude clouds occur 7,000–12,000 m (23,000–40,000 ft) above ground. They are typically small wispy or bobble-shaped clouds made up entirely of ice crystals. Multi-level cumulonimbus clouds (see pp.232–33) can tower through the entire height of the troposphere.

Veils and wisps visible at twilight, these extreme-level clouds form in the mesosphere

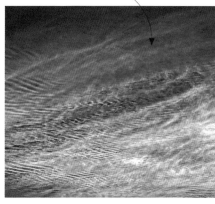

NOCTILUCENT

Very high-level stratospheric clouds visible at night and named for their resemblance to mother-of-pearl

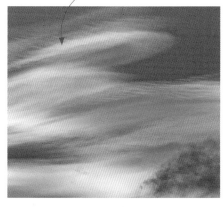

NACREOUS

Mid-level clouds

Lying 2,000–7,000 m (6,600–23,000 ft) above ground, mid-level clouds are dominated by common, sheetlike altostratus and by the patches, rolls, and ripples of altocumulus. More unusual types include the striking lenticular cloud shapes (altocumulus lenticularis) and the distinctive rounded pouches that can protrude in a cellular pattern from the underside of clouds, called mammatus.

Sheetlike patches and streaks at a relatively high level in the sky

ALTOSTRATUS

Dispersed, bulbous patches, sometimes arranged in long parallel bands

ALTOCUMULUS

Low-level clouds

On cloudy days, the whole lower troposphere, below 2,000 m (6,600 ft), can appear to be filled with low-lying, sheetlike stratus, fluffy cumulus, or mixed-form stratocumulus, as well as the bases of towering nimbostratus and cumulonimbus rainclouds. At this low level, the clouds are filled with water droplets. Other lenticular, tubular, and patchy clouds may also occur. Surface-level stratiform clouds are known as fog and mist.

Rare stratiform clouds with a rippling, wave-like underside, only added to the International Cloud Atlas in 2017

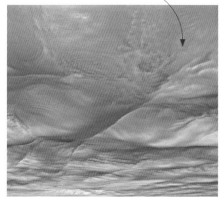

ASPERITAS

Rounded, bulbous clouds, sometimes arranged in parallel bands

STRATOCUMULUS

types of clouds

From iridescent wisps to bulbous pockets, the shapes, colours, and sizes of clouds vary wildly. Clouds form when moist air that has been warmed by the ground or driven over high ground rises upwards. As air rises it cools, reducing its ability to hold moisture. At a temperature known as the dew point, this air becomes saturated and the water vapour within it condenses around tiny particles called condensation nuclei, such as dust, pollen, and spores. This condensation forms minute droplets or freezes into tiny ice crystals and becomes the base of a cloud. When water condenses it releases latent heat, which allows the air to continue to rise and build up thick banks of cloud. The type of cloud created depends on temperature, humidity, and air stability.

Fine, wispy clouds associated with fine weather, but may portend a coming storm

CIRRUS

Lens-shaped clouds, often very smooth and with a bulbous top

ALTOCUMULUS LENTICULARIS

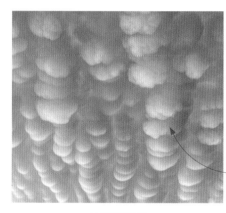

An unusual formation of globular pouches on the underside of a parent cloud

MAMMATUS

Fluffy, cotton-wool clouds that often appear in small patches

CUMULUS

Low-level, sheetlike clouds that may extend across the whole sky

STRATUS

Spectacular roll clouds formed by downdrafts from cumulonimbus thunderclouds

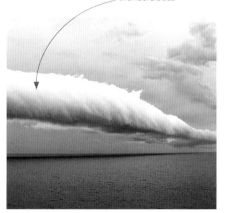

ARCUS

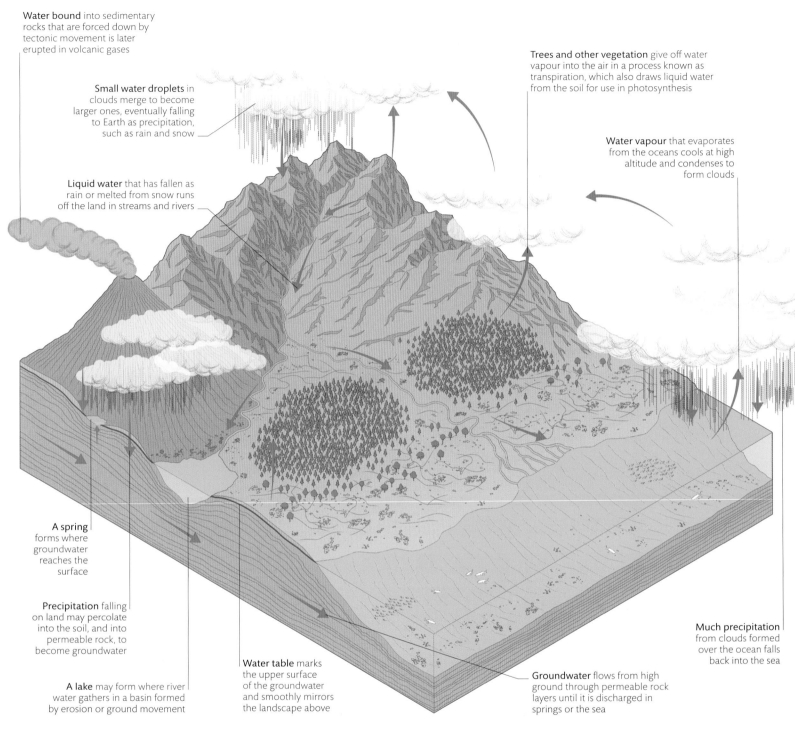

Water bound into sedimentary rocks that are forced down by tectonic movement is later erupted in volcanic gases

Small water droplets in clouds merge to become larger ones, eventually falling to Earth as precipitation, such as rain and snow

Liquid water that has fallen as rain or melted from snow runs off the land in streams and rivers

Trees and other vegetation give off water vapour into the air in a process known as transpiration, which also draws liquid water from the soil for use in photosynthesis

Water vapour that evaporates from the oceans cools at high altitude and condenses to form clouds

A spring forms where groundwater reaches the surface

Precipitation falling on land may percolate into the soil, and into permeable rock, to become groundwater

A lake may form where river water gathers in a basin formed by erosion or ground movement

Water table marks the upper surface of the groundwater and smoothly mirrors the landscape above

Groundwater flows from high ground through permeable rock layers until it is discharged in springs or the sea

Much precipitation from clouds formed over the ocean falls back into the sea

Perpetual transfer

Solar heat causes evaporation of water from oceans and land, and, by the process of transpiration, from plants. Water vapour rises, cools, and forms clouds. Eventually water in the atmosphere returns to Earth as precipitation. Melting of snow and ice, surface run-off in rivers, and infiltration into the ground all transfer liquid water from the land back to the oceans.

Trapped groundwater

Water from precipitation may seep though permeable rock into deeper rock layers, where it can become trapped by layers of impermeable rock. The pressure of water in the permeable rock may force the water out above ground, forming a natural spring. In some cases, humans can tap it to create an artesian well.

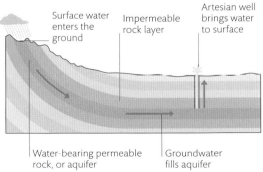

Surface water enters the ground

Impermeable rock layer

Artesian well brings water to surface

Water-bearing permeable rock, or aquifer

Groundwater fills aquifer

the water cycle

One of the most important systems that mould Earth's surface and regulate its atmosphere, the global water cycle (or hydrologic cycle) is driven ultimately by solar energy. It is vital for plant and animal life, drives the weather, and transfers energy through the oceans, atmosphere, and land. Water also lubricates plate movement and can trigger earthquakes. The total water volume of Earth is around 1.4 billion cubic km (333 million cubic miles), distributed on and below Earth's surface. The oceans hold nearly 96 per cent of this total, glaciers and ice sheets around 3 per cent, and subterranean groundwater about 1 per cent. All the world's rivers, lakes, atmosphere, and biosphere together contain a mere fraction (less than 1 per cent). A single water molecule may spend only a few days in the atmosphere or a few weeks in a river, but can be locked-up in an ice sheet for a million years.

Seasonal water cycle
Summer snowmelt in eastern Greenland feeds a steep torrent of water and sediment from the Syltoppene glacier across a rocky fan delta into King Oscar Fiord more than 2,700 m (8,900 ft) below. Heavy winter snowfall restarts this cycle.

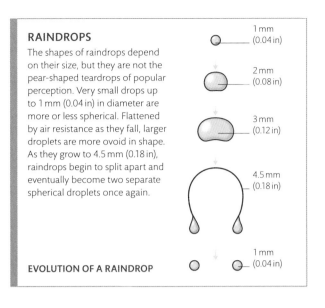

RAINDROPS

The shapes of raindrops depend on their size, but they are not the pear-shaped teardrops of popular perception. Very small drops up to 1 mm (0.04 in) in diameter are more or less spherical. Flattened by air resistance as they fall, larger droplets are more ovoid in shape. As they grow to 4.5 mm (0.18 in), raindrops begin to split apart and eventually become two separate spherical droplets once again.

1 mm (0.04 in)

2 mm (0.08 in)

3 mm (0.12 in)

4.5 mm (0.18 in)

1 mm (0.04 in)

EVOLUTION OF A RAINDROP

Downpour

A ship flees the deluge of a large cumulus congestus raincloud over the Black Sea in Europe. The seemingly impenetrable curtain beneath the cloud is in fact the downpour of falling rain spattering the sea surface. Where violent updrafts of air are too strong, or high temperatures cause evaporation, the rain may never reach Earth's surface.

falling rain

When water vapour in the atmosphere condenses around minute particles called condensation nuclei (see pp.222–23), the tiny droplets that form gather into clouds. The constant motion in clouds causes these droplets to coalesce and grow until they are heavy enough to fall to the ground. As the principal source of fresh water, rainfall is vital to life on Earth, and an unimaginable volume – over 500 trillion cubic metres (18,000 trillion cubic feet) – falls on the planet every year. Not just a valuable water source, rainwater also helps reduce greenhouse gas build-up, as carbon dioxide from the atmosphere dissolves in the water.

WINTER PRECIPITATION

At low temperatures, usually in winter, cold air causes snow instead of rain to form in the clouds. The type of precipitation that hits the ground, however, depends on the warm and cold air masses that it passes through on its way, as shown below. The red front depicts a warm air mass moving through cold air.

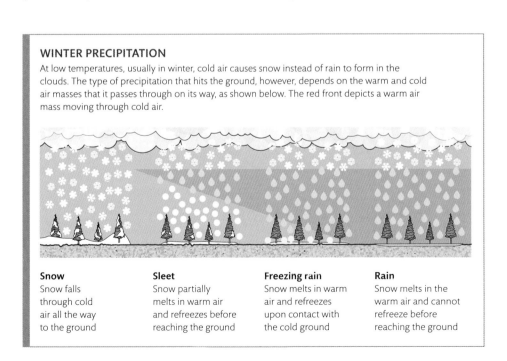

Snow
Snow falls through cold air all the way to the ground

Sleet
Snow partially melts in warm air and refreezes before reaching the ground

Freezing rain
Snow melts in warm air and refreezes upon contact with the cold ground

Rain
Snow melts in the warm air and cannot refreeze before reaching the ground

Sunseekers and surfers flock to the tropical Hawaiian Islands, US, in their millions each year, but, perhaps surprisingly, the islands boast one of the wettest places on the planet: the summit of Mount Wai'ale'ale on the island of Kauai. This extinct shield volcano rises to a height of 1,569 m (5,148 ft) and boasts lush green rainforest and boggy swampland nestled inside its deep crater, due to the high rainfall the mountain experiences.

Mount Wai'ale'ale

ocean and atmosphere

With a name that means "Rippling Water" or "Overflowing Water", Mount Wai'ale'ale has one of the highest rates of rainfall on Earth, typically ranging from 9.5 m to 11.4 m (31.2 ft to 37.4 ft) annually. It rains for between 335 and 360 days a year, and in 1982 alone, a staggering 17.3 m (56.8 ft) fell on the peak.

As the northernmost of the Hawaiian Islands, Kauai is surrounded by ocean and exposed to moisture-laden trade winds that bring rain from the northeast. Helped by the mountain's conical shape, incoming humid sea air is quickly driven up Wai'ale'ale's steep eastern walls, rising as high as 900 m (3,000 ft) over a distance of just 800 m (2,600 ft). As it rises and cools, this air condenses into clouds. Perhaps the most important factor that makes Wai'ale'ale one of the wettest spots not only in Hawaii, but also the world, is its height. The mountain's summit lies just below the trade wind inversion layer – the level at which trade winds cannot rise. This means that, unable to rise any further, the water-laden clouds are funnelled and squeezed into a small space and forced to drop a large portion of their rain on one concentrated spot: Mount Wai'ale'ale's peak.

Endemic to Mount Wai'ale'ale, this rare plant's compact, dome shape and dense whorls of leaves help it to thrive in the mountain's wet climate

Dubautia waialealae

Weeping Wall
Mount Wai'ale'ale's abundant precipitation feeds many waterfalls across the mountain, including these tumbling cascades that spill down a steep flank of the volcanic crater. One of the mountain's most famous sights, this lush, verdant wall is known as the Weeping Wall or Wall of Tears.

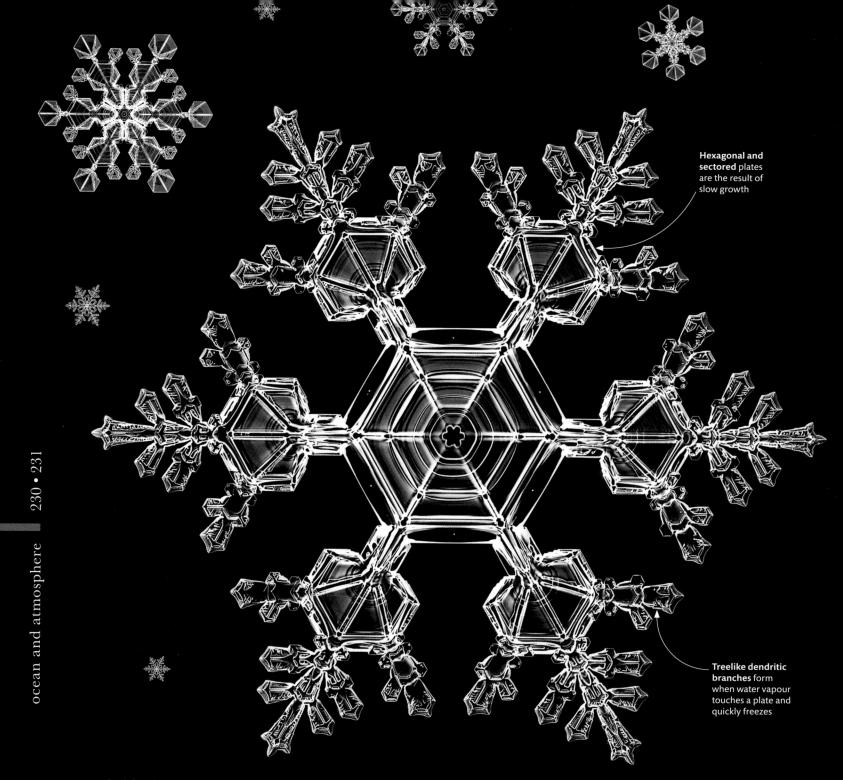

Hexagonal and sectored plates are the result of slow growth

Treelike dendritic branches form when water vapour touches a plate and quickly freezes

snow

In cold conditions, water vapour that condenses onto dust and other tiny particles can create minute ice crystals. Snowflakes grow as supercooled cloud droplets freeze onto these primary ice crystals, building them up into unique and often elaborate forms that are typically 2–10 mm (0.08–0.4 in) in size. Snowflakes are always hexagonal, due to the internal structure of the primary ice crystals they grow from. As they move through the turbulent air, they collide and stick together, gradually building up into the fluffy aggregate snowflakes that eventually land on the ground.

This snowflake has an almost entirely dendritic form, telling us it formed quickly throughout its growth

SNOWFLAKE MORPHOLOGY

The forms snowflakes take are partly dependent on the temperature and moisture content of the air. Solid plates and prisms form at low humidity, whereas more elaborate dendritic and delicate needle shapes form when humidity is higher. The largest flakes and heaviest snowfall occur between 0 °C (32 °F) and -5 °C (23 °F), when the air holds more moisture. At much colder temperatures the air is drier, resulting in less snow and smaller flakes.

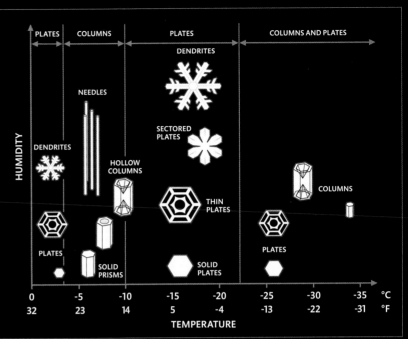

| PLATES | COLUMNS | PLATES | COLUMNS AND PLATES |

DENDRITES · NEEDLES · SECTORED PLATES · HOLLOW COLUMNS · DENDRITES · THIN PLATES · COLUMNS · PLATES · SOLID PRISMS · SOLID PLATES

HUMIDITY

0	-5	-10	-15	-20	-25	-30	-35	°C
32	23	14	5	-4	-13	-22	-31	°F

TEMPERATURE

Snowflakes

There is a magnificent array of delicate tracery in even the most common of snowflakes, as illustrated by these light micrographs. Slow growth of a snowflake can lead to hexagonal and sectored plates, while finer dendritic branches occur when growth is faster and less stable. Exposure to varied conditions can lead to snowflakes made up of several different forms.

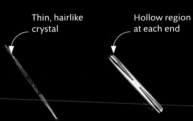

A very thin hexagonal prism

HEXAGONAL PLATE

A plate with broad arms and ridges

SECTORED PLATE

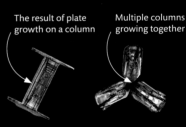

Thin, hairlike crystal

NEEDLE

Hollow region at each end

HOLLOW COLUMN

The result of plate growth on a column

CAPPED COLUMN

Multiple columns growing together

BULLET ROSETTE

A simple hexagonal plate forms the centre of this snowflake

Snowflake forms

Crystals of ice that are symmetrical and hexagonal, snowflakes' final forms depend on the conditions in which they grow. While dendritic shapes are most familiar, snowflakes may also take the form of plates, needles, columns, or rosettes.

Supercell
Sometimes the updrafts and downdrafts within a cumulonimbus separate and rotate around each other to form a more extensive storm front called a supercell. This supercell in Nebraska, US, sports forked lightning, a spiralling tornado, and extensive precipitation.

Hailstones
Made within the strong updrafts of cumulonimbus clouds, hailstones are rounded, solid ice balls formed by the freezing and accretion (accumulation of layers) of supercooled water.

Wet-growth hail forms when smaller ice-crystals collide and stick together

thunderstorms

Generated inside towering cumulonimbus clouds, thunderstorms have a dark menacing base that lies just a few hundred metres off the ground and multi-level, billowing plumes that can reach the top of the troposphere (the lowest layer of Earth's atmosphere) 12 km (7.5 miles) up. These storm clouds lead to sudden, brief deluges of rain, hail, or snow, accompanied by lightning flashes, thunder, and even tornadoes. They can store as much energy as 10 Hiroshima atom bombs, causing considerable damage when it is unleashed. Thunderstorms are most common in the warm, moist equatorial zone, but occur everywhere on Earth except for the polar regions.

FORMATION OF A THUNDERSTORM
When warm, moist air rises, it condenses into water droplets. If an air mass is highly unstable, this rise, called convection, can be rapid, leading to the formation of cumulonimbus clouds. Rapid convection releases heat, which causes further updrafts of warm air that can reach 160 km/h (100 mph), and associated downdrafts of cool air. If the cloud meets the tropopause, a flat, anvil-shaped top forms. Sometimes high-speed updrafts will break through the surface, creating an overshooting top.

Overshooting top — Storm motion

Downdraft

Tropopause (top of troposphere)

Anvil

Cumulonimbus cloud

Cool air

Updraft

Warm air

Precipitation

THUNDERSTORM STRUCTURE

The Great Plains of North America cut a broad swathe from Texas, US, to Canada. A land of weather extremes, these flatlands are home to Tornado Alley – a region famous for its high number of tornadoes. Although this region has more tornadoes than anywhere else in the world, scientists warn that the most intense, dangerous, and long-lived tornados often devastate regions to the south and east of the Great Plains.

spotlight Tornado Alley

A tornado begins near the top of a cumulonimbus cloud, where rising warm air starts to rotate slowly, creating a funnel shape in the centre. Downdrafts of cool air can force this spiralling funnel downwards, so that it protrudes from the base of the cloud. As the radius of the funnel decreases, it rotates more and more quickly, and when it touches the ground, it becomes a tornado. Tornadoes can last anywhere between a few seconds and over an hour. Meteorologists use the Enhanced Fujita scale to rate their intensity from EF0 to EF5. The fastest EF5 tornado winds ever recorded reached a staggering 480 km/h (300 mph). Tornadoes can span over 7 km (4.3 miles) in width and tear across land for more than 300 km (185 miles).

The loosely defined area of Tornado Alley, which sweeps across the US from Texas in the south to South Dakota in the north, records around 1,000 tornadoes each year. The region makes the ideal cauldron for the development of severe, tornado-spawning storms, as warm moist air travelling north from the Gulf of Mexico meets cold dry air from Canada and the Rocky Mountains, creating unstable atmospheric conditions.

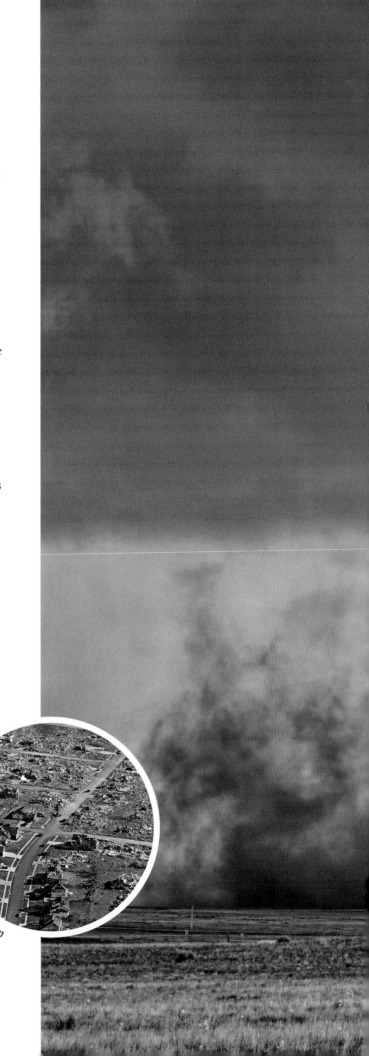

Complete devastation or near-miss is the lottery faced by residents of Tornado Alley

Path of destruction

Violent air
In the northern hemisphere, tornadoes typically rotate anticlockwise due to Earth's spin. Clockwise-rotating tornadoes, called anticyclonic tornadoes, are rare. This anticyclonic tornado near Simla, Colorado, US, has swept up many tons of dust and debris from the plains and is about to devastate a hapless ranch.

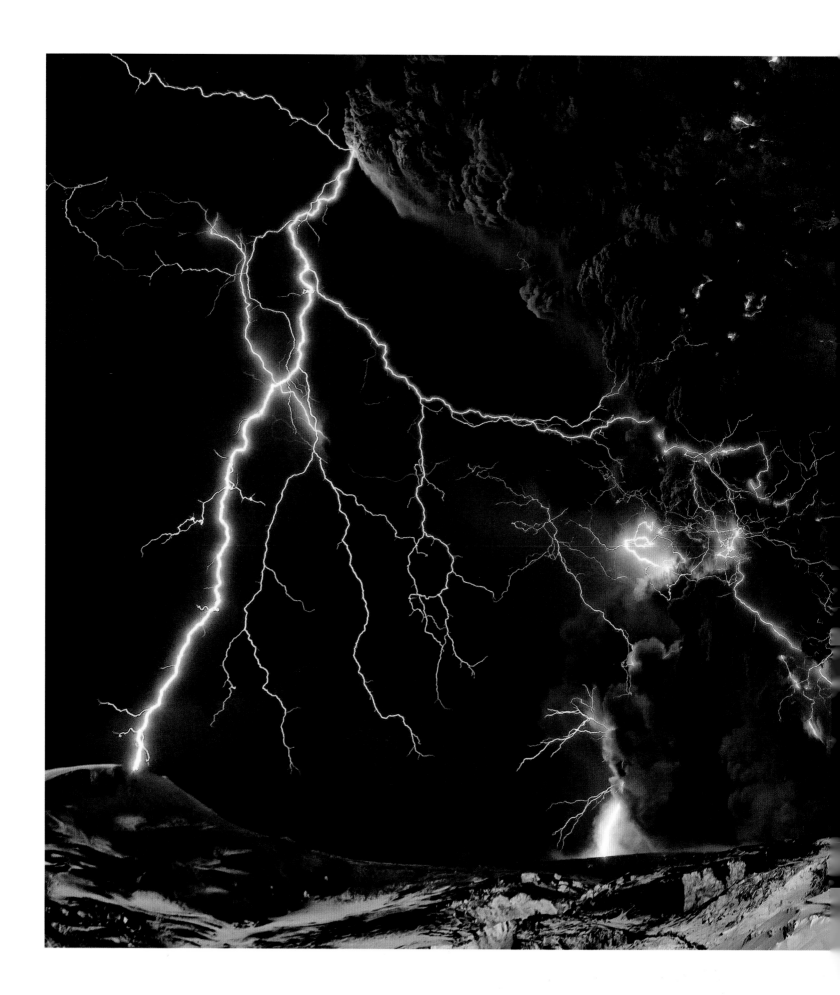

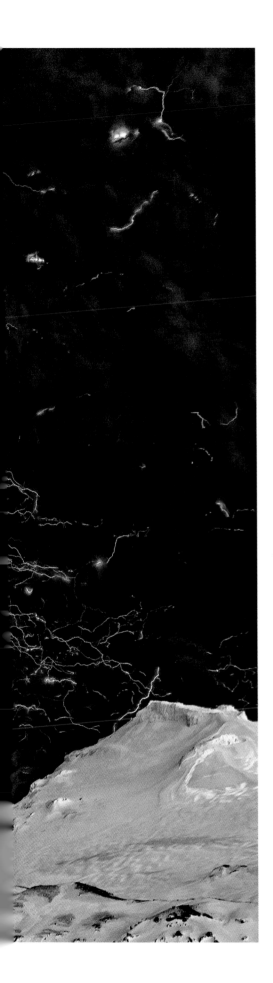

Blue colouration is caused by light scattering from many tiny spherules

Part of the soil has been fused by the intense heat

Natural glass from electricity
Fulgurites are a type of natural glass formed by the fusion of sand and silt under intense heat, such as when lightning strikes the ground. This blue fulgurite was created when a fallen tree channelled electricity from power lines into the ground over several hours, heating it and forming glass.

electricity in the air

On average there are three million lightning strikes worldwide per day, with about 70 per cent occurring over the tropics. Lightning is the dramatic result of a sudden discharge of static electricity from storm clouds. When static charge builds up an electric potential of several hundred million volts, the surrounding atmosphere can no longer insulate against its escape, leading to cloud-to-cloud lightning sheets or jagged lightning bolts that stretch to the ground. Typically two fingers wide and 2–3 km (1.2–1.8 miles) long, bolts emit dazzling light caused by superheating air to 30,000 °C (54,000 °F) in less than a second. This heat makes the air rapidly expand, resulting in a sonic shock wave, or thunderclap.

Volcanic lightning
During major eruptions – like this one from Iceland's Eyjafjallajökull volcano – ash, water, and gases are spewed high into the atmosphere. Turbulent movement of these particles within the ash cloud can lead to static electricity build-up and lightning, just as in rain clouds.

WHY LIGHTNING STRIKES

As they are pulled up and down by the air currents within a storm cloud, particles of rain, snow, ice, and soft hail (graupels) collide. Positive ions rub off particles and transfer to upward-moving smaller particles, such as ice crystals, while negative ions transfer to the heavier graupels that accumulate near the cloud base. A build-up of opposite charges eventually leads to lightning, within and between clouds, and directly to the ground below.

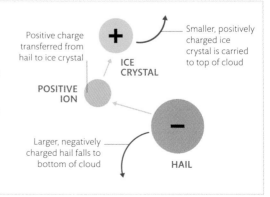

Positive charge transferred from hail to ice crystal

Smaller, positively charged ice crystal is carried to top of cloud

ICE CRYSTAL

POSITIVE ION

Larger, negatively charged hail falls to bottom of cloud

HAIL

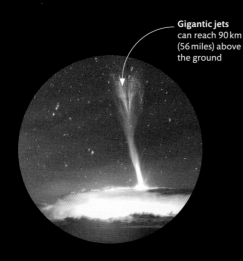

Gigantic jets
can reach 90 km
(56 miles) above
the ground

Red sprites

Fainter, faster, and significantly larger than lightning flashes, red sprites are elusive, short-lived optical events triggered far above active thunderstorms. Their red colour is due to light emitted from electrically ionized nitrogen. Their tendrils can become green–blue as they descend. These sprites were captured from the McDonald Observatory in Texas, US.

Gigantic jet

Seen here from the International Gemini Observatory in Hawaii, US, gigantic jets flare upwards from highly charged thunderclouds. The white and blue colours of a gigantic jet's initial burst end in a red fountain near the thermosphere (see p.210).

sprites, jets, and elves

Eyewitness accounts of pilots who glimpsed these unexplained lights were initially dismissed by meteorologists, until they were first caught on video in 1989. Part of a family of large-scale electrical discharge phenomena that occur high above major thunderstorms, called Transient Luminous Events (TLEs), red sprites, jets, and elves light up the atmosphere like fireworks. There are an estimated several million TLEs per year, but they are rarely witnessed from land, as they occur at altitudes of 40–100 km (25–60 miles) and mostly last only a millisecond. Sprites, which produce delicate hanging tendrils below a halo, are the most common.

TRANSIENT LUMINOUS EVENTS

TLEs are triggered by normal lightning discharge to the ground through the troposphere (see p.210), but mainly occur in the stratosphere and mesosphere as electrically induced forms of luminous plasma (charged particles). Sprites move downwards from the upper mesosphere, while the expanding red glows of elves occur higher still. Jets (cloud-to-air electrical discharges similar to lightning) extend tens of kilometres upwards. Other TLEs, including trolls, pixies, ghosts, and gnomes, are still poorly understood.

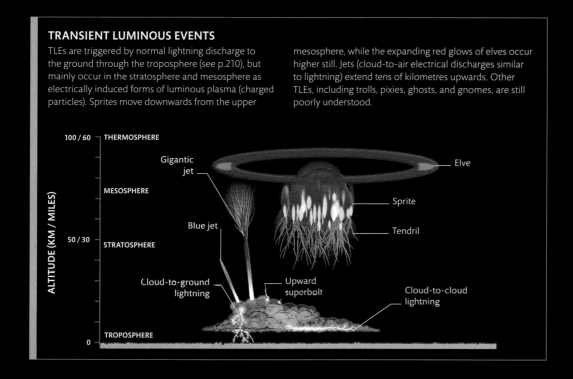

ALTITUDE (KM / MILES)

100 / 60 — THERMOSPHERE

MESOSPHERE

Gigantic jet

Elve

Sprite

Tendril

50 / 30 — STRATOSPHERE

Blue jet

Cloud-to-ground lightning

Upward superbolt

Cloud-to-cloud lightning

0 — TROPOSPHERE

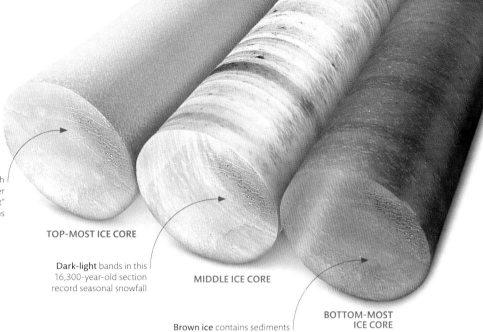

Ice cores from Greenland

Narrow cylinders of ice are extracted by rotating drills, or thermal drills, which melt the ice. The cores are stored in giant freezers and then examined and sampled. Here, three sections extracted from the Greenland ice sheet contain information revealed by ice from different depths.

Compacted ice – from a depth of about 53 m (175 ft), and over 170 years old – reveals "recent" snowfall patterns

TOP-MOST ICE CORE

Dark-light bands in this 16,300-year-old section record seasonal snowfall

MIDDLE ICE CORE

Brown ice contains sediments and volcanic ash from more than 111,000 years ago

BOTTOM-MOST ICE CORE

history of Earth science

ice-core analysis

Ice cores contain tiny bubbles of atmospheric gases from the distant past that have been trapped in ice and buried beneath glaciers and ice sheets, and yield a climate record over a period of up to 800,000 years. A few cores contain ice of more than 2 million years old. The deepest cores from Antarctica penetrate over 3,000 m (10,000 ft) and contain clear evidence of the correlation between levels of greenhouse gases and climate change.

Researching the Greenland ice sheet

Ernst Sorge (1899–1946), a German glaciologist, is pictured here during an expedition to Greenland in 1930–31. On this mission, he dug a shaft 15 m (49 ft) deep to make the first records of the density and temperature of buried ice.

Scientists have long been curious to discover the secrets that lie under the thick layers of ice that occur in glaciers and the polar regions. Louis Agassiz (1807–73), a Swiss-born American pioneer of theories about ice ages, used iron rods to drill into Alpine glaciers. Open-hole drilling, vertical shafts, and trenching were used to investigate the Antarctic ice in the early 20th century. Similar methods were used by Ernst Sorge (see inset, above left) for his research into the Greenland ice sheet in the 1930s.

The first ice cores (cylinders of ice) drilled from deep in the ice layer were recovered from Greenland and Antarctica in the 1950s. Since then, large numbers of ice cores have been drilled and stored for research. Scientists can detect the levels of key elements in these ice cores to date the samples accurately. Different forms of oxygen occur in ice formed during and between glacial periods. Such evidence helps to give a precise record of past climates, including snow accumulation, temperature changes, glacier retreat, and variation in ocean and atmospheric composition. Particulate matter in ice cores can also provide evidence of major volcanic eruptions and wild fires.

Analysis of ice cores from glaciers across the world is also used to further advance our understanding of natural climate variation.

Air bubbles from the past

This ultra-thin sliver of Antarctic ice shows hundreds of tiny air bubbles trapped between ice crystals. The bubbles contain nitrogen, oxygen, argon, carbon dioxide, and methane, and provide vital data on the composition of the atmosphere at the time they became trapped. Polarized lighting of the ice causes the colourful effect.

> " The era of intense cold, which preceded the present … has been only a temporary oscillation of the Earth's temperature. "
>
> LOUIS AGASSIZ, 1837

The current ice age

During icehouse periods, or ice ages, Earth cycles between colder periods (glaciations), with extensive polar icecaps, and warmer interglacials, with less ice and higher seas. Today we live in an interglacial of the current icehouse period. The ice has receded since the last glaciation, which peaked 22,000 years ago. Even so, the Antarctic icecap is 2,100 m (6,900 ft) thick. It buries mountains, leaving just their peaks visible as so-called "nunataks".

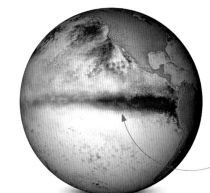

El Niño events

Every few years, a natural periodic oscillation of the ocean–atmosphere system significantly warms Pacific surface waters. Known as El Niño, the phase profoundly affects Earth's climate and can warm the planet by 0.7 °C (1.3 °F).

The surface water of the eastern Pacific experiences excess warmth (red), leading to drier, hotter weather across North America

natural climate change

About 100 million years ago, palm trees and crocodiles lived near the North Pole; now the region is covered with thick ice. Earth's climate cycles between two states: greenhouse and icehouse Earth. Icehouse periods, also known as ice ages, feature lower temperatures, continental ice sheets, and icecaps in the polar regions, whereas greenhouse periods experience an increase in global temperatures and the retreat of ice. These natural changes to the planet's climate are driven by variations in solar radiation, the position of continents and oceans, and greenhouse gases.

MILANKOVITCH CYCLES

Fluctuations in Earth's orbit around the Sun and rotation on its axis lead to additional cyclical changes in the climate. Named after Milutin Milankovitch, the Serbian mathematician who first described their effect, these cycles are responsible for the alternating glacial and interglacial conditions during the current icehouse period. Also known as the Quaternary Ice Age, this began just over 2.5 million years ago.

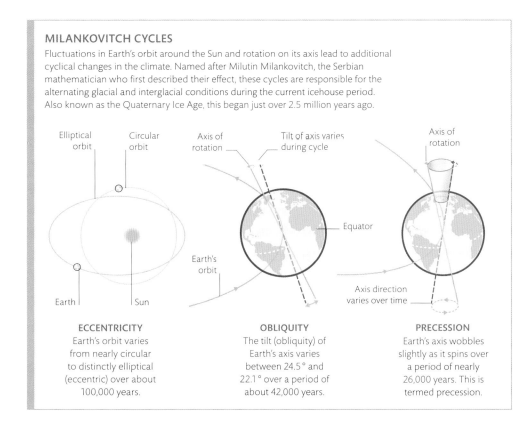

ECCENTRICITY
Earth's orbit varies from nearly circular to distinctly elliptical (eccentric) over about 100,000 years.

OBLIQUITY
The tilt (obliquity) of Earth's axis varies between 24.5° and 22.1° over a period of about 42,000 years.

PRECESSION
Earth's axis wobbles slightly as it spins over a period of nearly 26,000 years. This is termed precession.

living planet

The living part of Earth, its biosphere, forms a thin film on its surface measuring less than two-thousandths of the planet's diameter. This veneer has nevertheless had a profound influence on the planet's development over the last 4 billion years. Living processes have transformed Earth's atmosphere, and living organisms have left a fossil record that allows us to trace the path of evolution and to reconstruct geological history in fine detail.

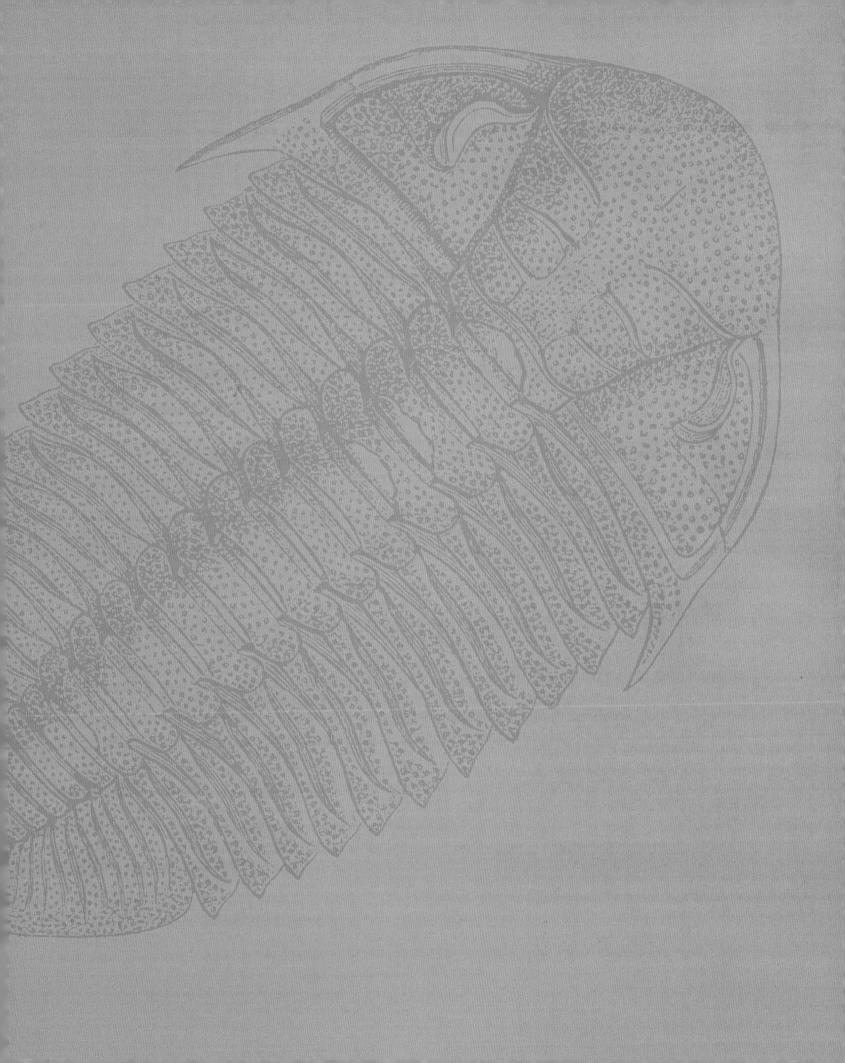

the biosphere

The parts of Earth that are habitable by life are confined to its surface – in a layer of rock, water, and air called the biosphere. The thickness of this layer accounts for less than 0.2 per cent of the diameter of the planet. This extraordinary living coating is the only example yet discovered in the universe. At its deepest, it follows ocean trenches that plunge as far down into the crust as the highest mountains that rise above it. At these extremes, life is sparse and specialized to cope with hostile conditions. But closer to sea level – on land and in the oceans – life teems in forests, deserts, rivers, lakes, and the vast marine realm. Almost all of this life is fuelled by sunlight, as food-making plants and algae nourish complex food chains.

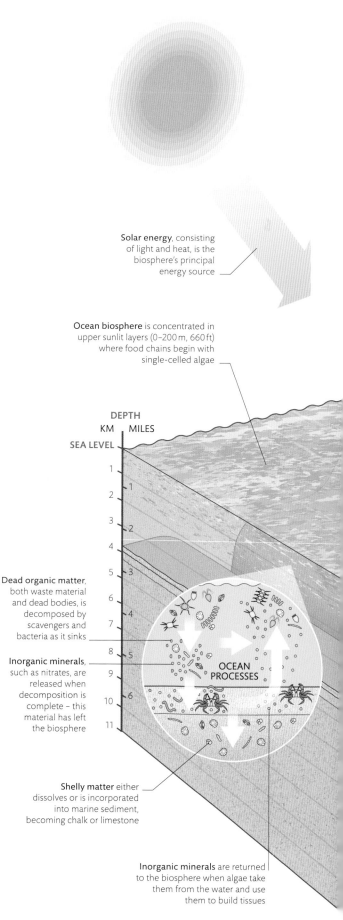

Solar energy, consisting of light and heat, is the biosphere's principal energy source

Ocean biosphere is concentrated in upper sunlit layers (0–200 m, 660 ft) where food chains begin with single-celled algae

DEPTH
KM   MILES
SEA LEVEL

Dead organic matter, both waste material and dead bodies, is decomposed by scavengers and bacteria as it sinks

Inorganic minerals, such as nitrates, are released when decomposition is complete – this material has left the biosphere

OCEAN PROCESSES

Shelly matter either dissolves or is incorporated into marine sediment, becoming chalk or limestone

Inorganic minerals are returned to the biosphere when algae take them from the water and use them to build tissues

Life fuelled by minerals, not sunlight
In parts of the deep ocean floor, life flourishes without the Sun's energy. Here, volcanic hydrothermal vents emit hot streams of mineral-rich sea water. Bacteria release energy from the minerals to make food, while invertebrates graze on these bacteria and are, in turn, food for predators, such as fish.

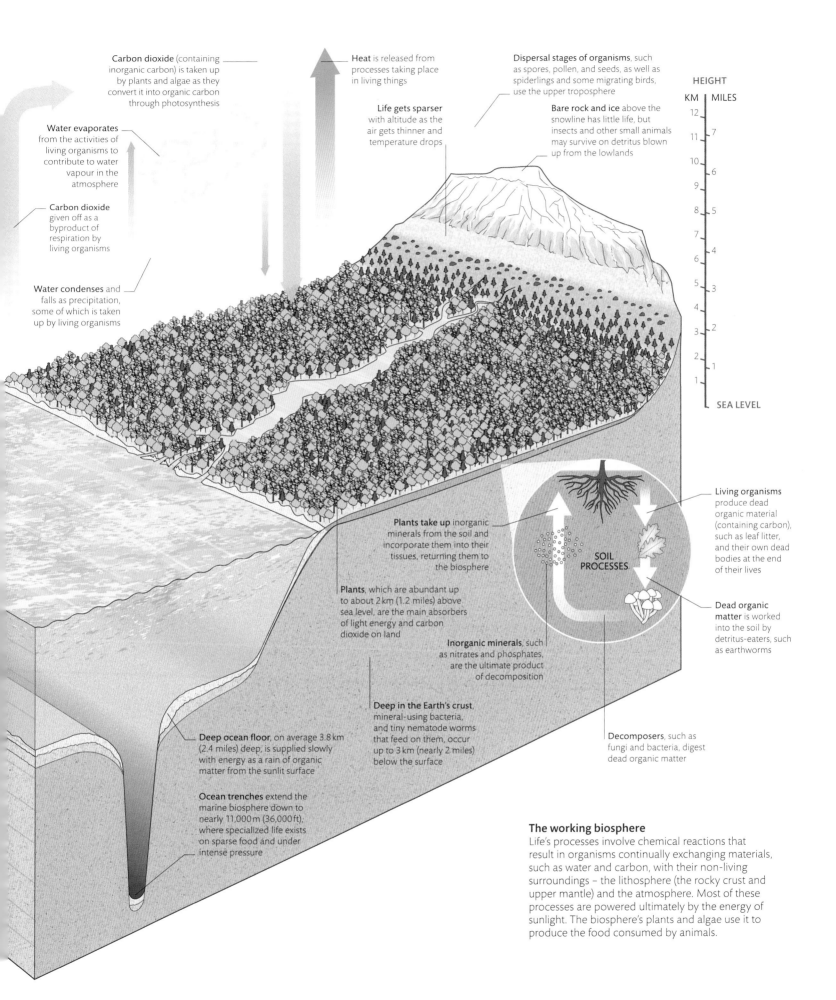

Carbon dioxide (containing inorganic carbon) is taken up by plants and algae as they convert it into organic carbon through photosynthesis

Heat is released from processes taking place in living things

Dispersal stages of organisms, such as spores, pollen, and seeds, as well as spiderlings and some migrating birds, use the upper troposphere

Water evaporates from the activities of living organisms to contribute to water vapour in the atmosphere

Life gets sparser with altitude as the air gets thinner and temperature drops

Bare rock and ice above the snowline has little life, but insects and other small animals may survive on detritus blown up from the lowlands

HEIGHT

KM | MILES

12

11 — 7

10 — 6

9

8 — 5

7

6 — 4

5 — 3

4

3 — 2

2

1 — 1

SEA LEVEL

Carbon dioxide given off as a byproduct of respiration by living organisms

Water condenses and falls as precipitation, some of which is taken up by living organisms

Plants take up inorganic minerals from the soil and incorporate them into their tissues, returning them to the biosphere

Living organisms produce dead organic material (containing carbon), such as leaf litter, and their own dead bodies at the end of their lives

SOIL PROCESSES

Plants, which are abundant up to about 2 km (1.2 miles) above sea level, are the main absorbers of light energy and carbon dioxide on land

Dead organic matter is worked into the soil by detritus-eaters, such as earthworms

Inorganic minerals, such as nitrates and phosphates, are the ultimate product of decomposition

Deep in the Earth's crust, mineral-using bacteria, and tiny nematode worms that feed on them, occur up to 3 km (nearly 2 miles) below the surface

Deep ocean floor, on average 3.8 km (2.4 miles) deep, is supplied slowly with energy as a rain of organic matter from the sunlit surface

Decomposers, such as fungi and bacteria, digest dead organic matter

Ocean trenches extend the marine biosphere down to nearly 11,000 m (36,000 ft), where specialized life exists on sparse food and under intense pressure

The working biosphere
Life's processes involve chemical reactions that result in organisms continually exchanging materials, such as water and carbon, with their non-living surroundings – the lithosphere (the rocky crust and upper mantle) and the atmosphere. Most of these processes are powered ultimately by the energy of sunlight. The biosphere's plants and algae use it to produce the food consumed by animals.

millions of species

Earth is the only planet known to have life and there is a breathtaking variety of colour, shape, and movement in its surface biosphere (see pp.246–47). Every species has developed its own way of surviving and reproducing, and each one is descended by evolution from a single ancestor that was just a cell some 4 billion years ago. Today, the diversity of microbes, plants, and animals is at its richest in ancient forests and sun-drenched coral reefs, but life forms can exist anywhere they can feed and take in oxygen – and have found a way to do so – from the highest mountains to the deepest ocean trenches.

Pollination partners

Insects and flowering plants are the most species-rich elements of Earth's biodiversity. As these marbled white butterflies, *Melanargia galathea*, drink nectar from greater knapweed, *Centaurea scabiosa*, and transfer pollen, they continue a partnership of insects and flowering plants that evolved millions of years ago and led to the planet's astonishing diversity of both.

Prehistoric diversity

This 420-million-year-old slab of rock is a snapshot of early biodiversity. It carries fossils of a reef community, ancestors of modern marine animals.

Rugose coral is a relative of modern stony coral (p.259)

Fragment of a fan-shaped colony of *Fenestella* – tiny filter-feeding animals known as bryozoans

HOW SPECIES EVOLVE THROUGH GEOGRAPHICAL SEPARATION

Over many millions of years, life on Earth changes through a process of evolution. As this happens, populations split and their characteristics can diverge so much that they become different species. The diversity of species at any one time comes from a balance between new species generated and those that become extinct because their populations do not survive.

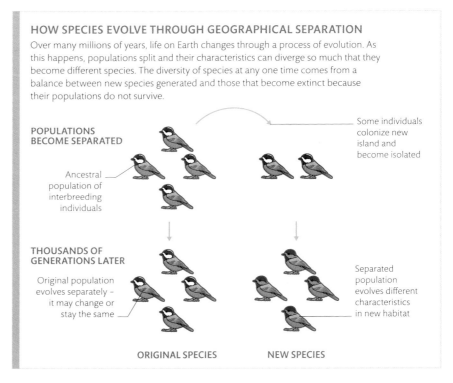

POPULATIONS BECOME SEPARATED

Ancestral population of interbreeding individuals

Some individuals colonize new island and become isolated

THOUSANDS OF GENERATIONS LATER

Original population evolves separately – it may change or stay the same

Separated population evolves different characteristics in new habitat

ORIGINAL SPECIES   NEW SPECIES

early life

For the first billion years of its existence, Earth was lifeless. But given the vast timescale available, the rare chemical reactions that were possible on its mineral-rich surface became likely – perhaps inevitable. In turn, these reactions produced the first complex organic matter, and eventually, self-replicating molecules enveloped in oily membranes became the first living cells. No-one knows exactly where the first life appeared, although a strong theory points to volcanic vents on the ocean floor. But by at least 3.4 billion years ago, life had colonized shallow seas, leaving strange fossilized rocky mounds called stromatolites.

Small, dark vertical shapes are impressions left by algal tufts that grew towards the sunlight

Side "branches" arose as the single-celled algae moved upwards on a layer of slime

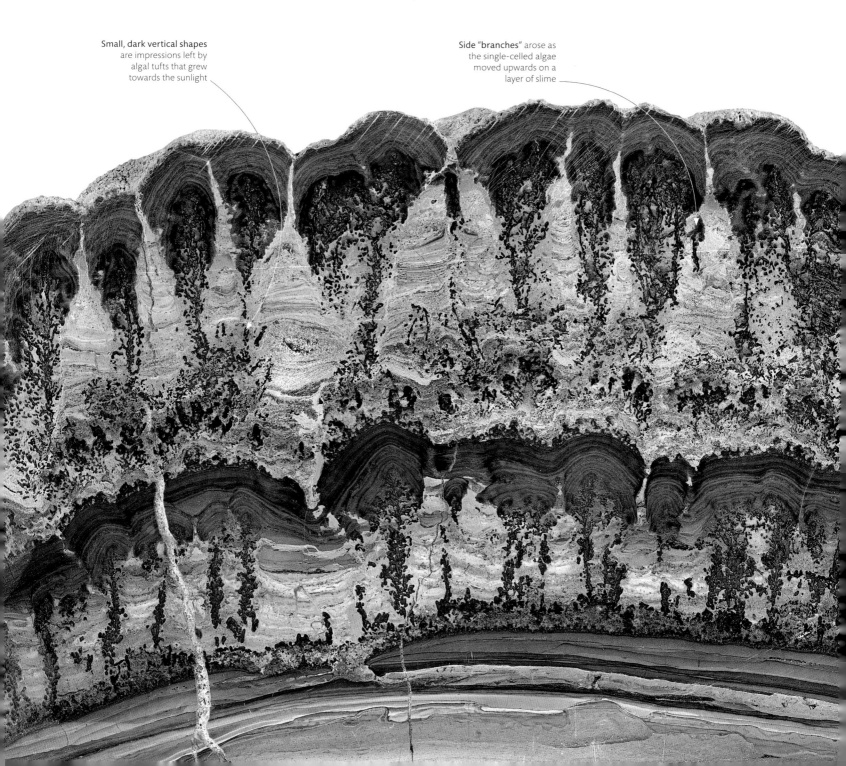

Living stromatolite
Stromatolites, like these in Shark Bay, Western Australia, still grow today. They are confined to places that are inhospitable to animals that graze the microbes that produce them, such as areas where the water is too salty.

Active microbes remain on the surface layer

Lower section is formed by remnants of former microbial surface communities

Dome-shaped layers built up over remains of algal tufts

Lower layers formed from algae that died as new algae grew above them, blocking out the light

FROM BIOFILM TO STROMATOLITE

Even today, many kinds of microbes – including bacteria and algae – produce sticky slime that keeps their cells amassed into thin colonies, or biofilms. Photosynthetic microbes called cyanobacteria need light, so continually migrate upwards, leaving dead layers of the microfilm beneath. The slime binds sediment, which accumulates in layers that build up over thousands of years and cement together to form rocky stromatolites.

Photosynthetic cyanobacteria

Slime film secreted by cyanobacteria binds sediment

STAGE 1

New cyanobacteria form and migrate towards sunlight

Dead cyanobacteria become entombed in slime and sediment

STAGE 2

Cyanobacteria continue to move towards light

Layers of sediment and old cyanobacteria (microbial film) build

STAGE 3

Layers of dead microbial films form rock

Photosynthetic cyanobacteria at the surface

MATURE STROMATOLITE

From microbes to marble

This slab of Cothan marble from the dawn of the dinosaurs – billions of years after the first stromatolites formed – shows that simple microbes have been sculpting rocks throughout life on Earth. The dark, blue-black bushy streaks were once sticky tufts of living algae that bound mounds of mud; over thousands of years, they solidified into limestone, then crystallized to marble.

Banded iron formation

Iron ore develops from iron oxides – the products of the reaction between soluble iron salts and oxygen. As oxygen-releasing microbes waxed and waned over time, bands were formed in sedimentary rocks on the seabed; the bands correspond to the seasonal changes in the activity of microbes. Earth's shifting crust has brought the rocks, such as this one found in the iron mining districts of Western Australia, to the surface.

HOW PREHISTORIC LIFE OXYGENATED THE WORLD

The first photosynthesizing organisms were microbes that build rocky mounds called stromatolites (see pp.250–51) found in sunlit shallow seas. Iron salts welling up from the deep ocean reacted with the oxygen to deposit bands of solid iron oxide, or rust. As iron salts in the water declined, the oxygen from photosynthesis effectively overspilled into the atmosphere, eventually rising to the 21-per-cent-level of today.

Oxygen in sea water from photosynthesis

Photosynthesizing microbes form a mat on stromatolite and give off oxygen

Atmosphere

Iron salt reacts with oxygen, forming solid iron oxide

Iron oxide (rust) settles on sea floor

Iron salts in sea water

Rock

MICROBES RELEASE OXYGEN

Atmosphere

Iron salt

Ocean

Oxygen

Layer of iron oxide accumulates

Rock

IRON OXIDE LAID DOWN

Oxygen escapes into atmosphere

Atmosphere

Ocean

Iron salts depleted

Band of iron forms in rock

Rock

AIR FILLS WITH OXYGEN

Lighter coloured bands are composed mainly of scattered grains of chalcedony, a silica-based mineral containing iron oxide

Tiny cells of *Nostoc* form chains linked by branchlike structures

Darkest bands have the richest concentration of iron and are made of iron oxide minerals called magnetite and haematite

Bubbling oxygen
Slimy underwater colonies of modern bacteria, called *Nostoc*, produce bubbles of oxygen. Today these photosynthesizers – along with more complex algae and plants – keep air and water oxygen-rich.

Golden bands, known as "tiger eye", carry an intermediate amount of iron

life transforms Earth

More than 2 billion years ago, the first green photosynthetic microbes bubbled up oxygen, producing some of Earth's most striking rock patterns, and affecting the course of evolution forever. Some of Earth's richest concentrations of iron ore are the result of this event at the dawn of life. The oxygen reacted with dissolved iron particles floating in sea water, producing minerals deposited under prehistoric oceans as solid red ore. When the soluble iron in the oceans was depleted, the oxygen from the microbes escaped into the air – helping to create the breathable atmosphere that life relies on today.

life branches out

As the first single-celled organisms evolved into bigger, multi-celled bodies, life began to have a bigger physical impact on the world around it. Animals have nerves and muscles so their reactions and movements are faster and strong enough to churn sediment and stir water as they search for nutrients in the water column, or from dead matter on the ocean beds. The first simple animals flourished on the seafloor around 600 million years ago, during a period called the Ediacaran, named after the Ediacara Hills in South Australia, where their fossils are especially abundant.

Wide head had sense organs concentrated in the animal's direction of movement

Forward movement
Just 3–5 cm (1½–2 in) long, *Spriggina* was a complex animal with distinctively different ends to its body. It must have moved in one direction – head first.

HOW ORGANISMS CHANGED THE SEABED

The movements of the first muscle-powered animals changed conditions beyond the flat seabed, making other parts of the ocean more habitable for life. Some animals bulldozed the surface, while others burrowed down. Together they stirred up ocean sediment, which helped oxygenate the mud below and nourish the waters above.

CROSS-SECTION OF SEAFLOOR

Fixed, upright organisms absorbed nutrients from water

Wormlike animals worked into sediment

Burrowing animals dispersed nutrients into the water

Photosynthetic microbial mats covered the seafloor

Life on the seafloor

The ribbed, leaflike impressions on these *Dickinsonia* fossils indicate they were organisms that lived on the seafloor. The prominent ribs suggest that their bodies were supported by a tough hornlike material. Trace fossils on the rock could be track marks, evidence that they moved from place to place.

Feathery
"arms" trap
food particles

CRINOID
Order Comatulida

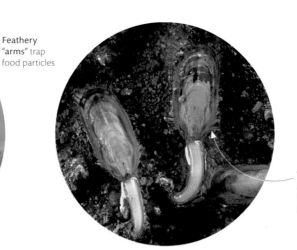

Two-part shell
similar to that of
bivalve molluscs,
such as mussels

BRACHIOPOD
Lingula hians

Cambrian origins
The Cambrian explosion was a time of intensive evolution, when many animal body plans originated. Today's crinoids – relatives of starfish – are very similar to their Cambrian counterparts. Some animals, such as brachiopods, have changed so little in half a billion years that they are described as "living fossils".

Cambrian explosion

As the first wriggling animals churned mud and used ocean minerals to build the first shells, life began to affect the cycles of sediment and rocks beyond the living biosphere. Shell-building and reef-forming animals amassed calcium and silica in their bodies, and many of them used muscle power to swim upwards into the water column. Early in a period of time called the Cambrian, about 539–520 million years ago, an explosion in the development of these hard-shelled animals found the prehistoric seas teeming with the distant ancestors of sea creatures alive today – including shrimplike arthropods, snails, and sea urchins.

Shelly fauna
Exoskeletons and shells preserve better than soft bodies, and it is the predominance of shelled fauna from the Cambrian that helps to explain the rich diversity of its fossil record. Seen here are the conelike tubes of *Archotuba* (centre) – which may have enclosed a soft, tentacled anemone – and a jointed-legged arthropod (top), called a trilobite (see pp.278–79).

CAMBRIAN OCEAN DIVERSITY

Fossils of the ocean-dwelling animals of the Cambrian reveal body components, such as mouthparts and limbs, that are important clues not only about their feeding habits but also about how they moved. Their diversity points to a community of swimmers and crawlers, scavengers and predators, which were all linked in the first complex food chains. By dispersing upwards from the ocean floor, where the first animals originated, they colonized the open water – some drifting in the currents as the earliest plankton, while others swam against the currents. In this way, animal life expanded the biosphere into three dimensions of the ocean.

THE CAMBRIAN OCEAN FOOD CHAIN

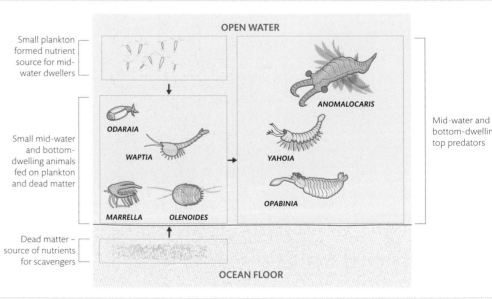

OPEN WATER

Small plankton formed nutrient source for mid-water dwellers

Small mid-water and bottom-dwelling animals fed on plankton and dead matter

Dead matter – source of nutrients for scavengers

ODARAIA

WAPTIA

MARRELLA   *OLENOIDES*

ANOMALOCARIS

YAHOIA

OPABINIA

Mid-water and bottom-dwelling top predators

OCEAN FLOOR

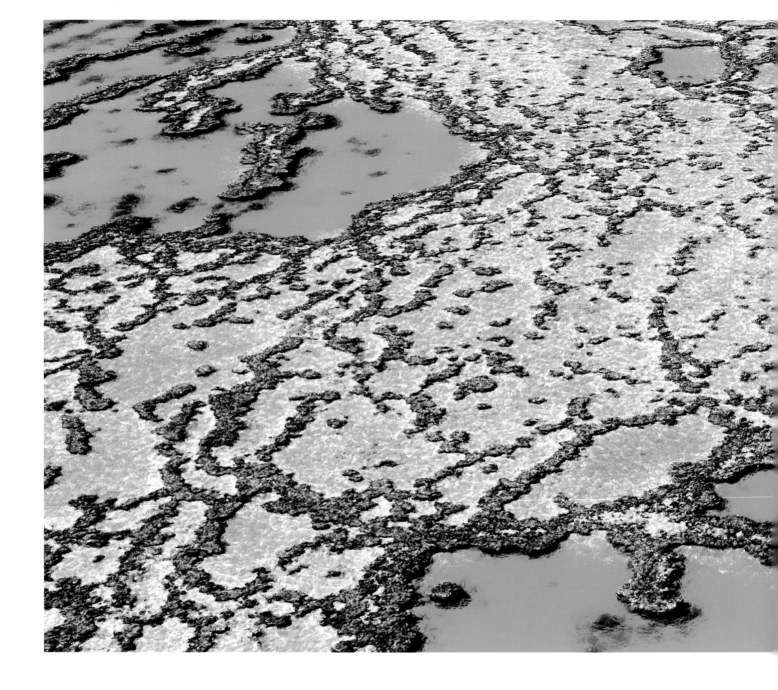

making reefs

Ocean reefs are the biggest structures made by living things and can be seen from space. They are constructed by organisms depositing sedimentary rock over millennia. Today's reefs are built by coral polyps using minerals extracted from sea water. These colonial animals are cousins of anemones and jellyfish – a group whose ancestry stretches back half a billion years. But the history of prehistoric reefs is far richer – once stromatolites (see pp.250–51) were grazed almost to oblivion, organisms as diverse as microbes and sponges and, later, rugose corals, succeeded them as the main reef-builders.

Great Barrier Reef
Stretching halfway down Australia's east coast, the Great Barrier Reef is Earth's largest modern reef system. Its complex geography of underwater ridges and islands creates a mosaic of underwater habitats. Like reefs of the prehistoric past, it is a "hotspot" of marine biodiversity: a place where huge numbers of species colonize and evolve.

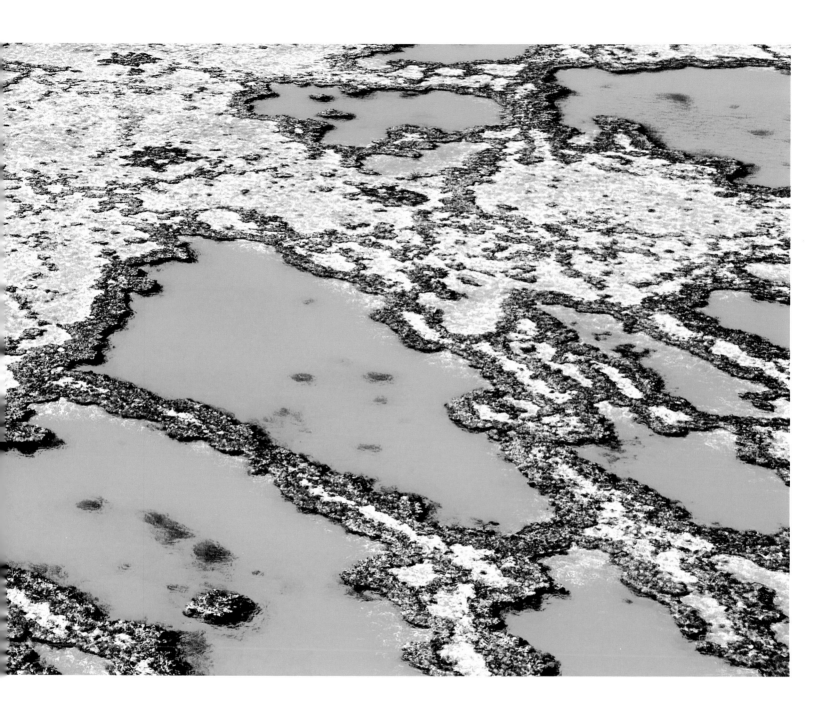

Fossilized reef

Some of Earth's earliest reefs were made during the Cambrian period by animals called archaeocyathid sponges. Their rocky colonies were built from a series of hollow cones.

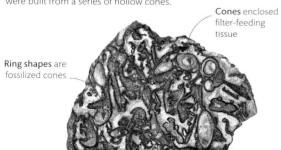

Cones enclosed filter-feeding tissue

Ring shapes are fossilized cones

REEF BUILDERS, PAST AND PRESENT

Modern reef-forming corals have been around since the time of the dinosaurs, but earlier reefs were built mainly by different kinds of organisms. The first were made from microbial films or sponges. Most of these were later replaced by hornlike rugose corals and then by modern stony corals.

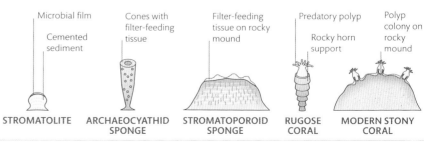

STROMATOLITE	ARCHAEOCYATHID SPONGE	STROMATOPOROID SPONGE	RUGOSE CORAL	MODERN STONY CORAL
Microbial film	Cones with filter-feeding tissue	Filter-feeding tissue on rocky mound	Predatory polyp	Polyp colony on rocky mound
Cemented sediment			Rocky horn support	

THE DEVELOPING FISH

Vertebrates have a skeleton made of cartilage and bone. Their fishlike invertebrate ancestors, from more than 500 million years ago, had neither – but their body had a rubbery rod, called a notochord, which ran through their back and helped to strengthen their side-to-side swimming movements. Over time, the notochord was reinforced with cartilage and bone that also enclosed and protected the brain and spinal cord, as well as bracing the gills. The gill braces, or arches, kept the gills open for absorbing oxygen. Later, the front gill arches evolved into hinged jaws, which helped to manipulate and bite food.

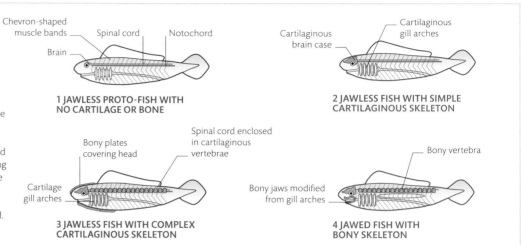

1 JAWLESS PROTO-FISH WITH NO CARTILAGE OR BONE
- Chevron-shaped muscle bands
- Brain
- Spinal cord
- Notochord

2 JAWLESS FISH WITH SIMPLE CARTILAGINOUS SKELETON
- Cartilaginous brain case
- Cartilaginous gill arches

3 JAWLESS FISH WITH COMPLEX CARTILAGINOUS SKELETON
- Bony plates covering head
- Spinal cord enclosed in cartilaginous vertebrae
- Cartilage gill arches

4 JAWED FISH WITH BONY SKELETON
- Bony vertebra
- Bony jaws modified from gill arches

Wide head formed part of armourlike bony exoskeleton

Circular openings, or orbits, in the head shield for the eyes

Jawless fish
The earliest fish – such as ostracoderms – lacked jaws and probably grubbed around near the ocean floor. This *Zenaspis* from the early Devonian lived in shallow seas and river mouths.

Giant predator
This fossil of *Dunkleosteus*, a giant armour-plated fish from the late Devonian, shows that by around 400 million years ago, backboned animals had evolved into the biggest biting species of the time – something that made them top predators of increasingly complex food chains. *Dunkleosteus* belongs to a group of fish, called placoderms, that were the earliest vertebrates with jaws.

age of fish

As living things filled Earth's waters, animals evolved to be bigger, faster, and stronger, especially among the first backboned animals (vertebrates) – the ocean's fish. Their skeletons supported growing bodies and in some groups became protective armour. Fish had descended from small soft-bodied invertebrates that used gills to filter nourishment, but now sucked larger morsels of food into their mouths with their muscular throat. This left their gills to focus on extracting oxygen from the water. By the Devonian period, 400 million years ago, the oceans were filled with numerous groups of fish. In some, the gills' supporting rods had evolved into another key innovation: biting jaws.

Thick fleshy fins moved fish though water

Paddle-shaped pectoral fin

Fleshy finned jawed fish
Tristichopterus was a late Devonian fish more closely related to living jawed vertebrates than placoderms. Some of its later cousins evolved walking limbs and became land vertebrates.

Jaw's jagged edge
provided cutting surface
similar to the teeth that
evolved in later jawed
fish, like sharks

Hinged neck joint allowed
head to pivot upwards,
helping to widen mouth
before biting

Joint between upper and
lower jaw was worked by
strong muscles that could
deliver powerful bite force

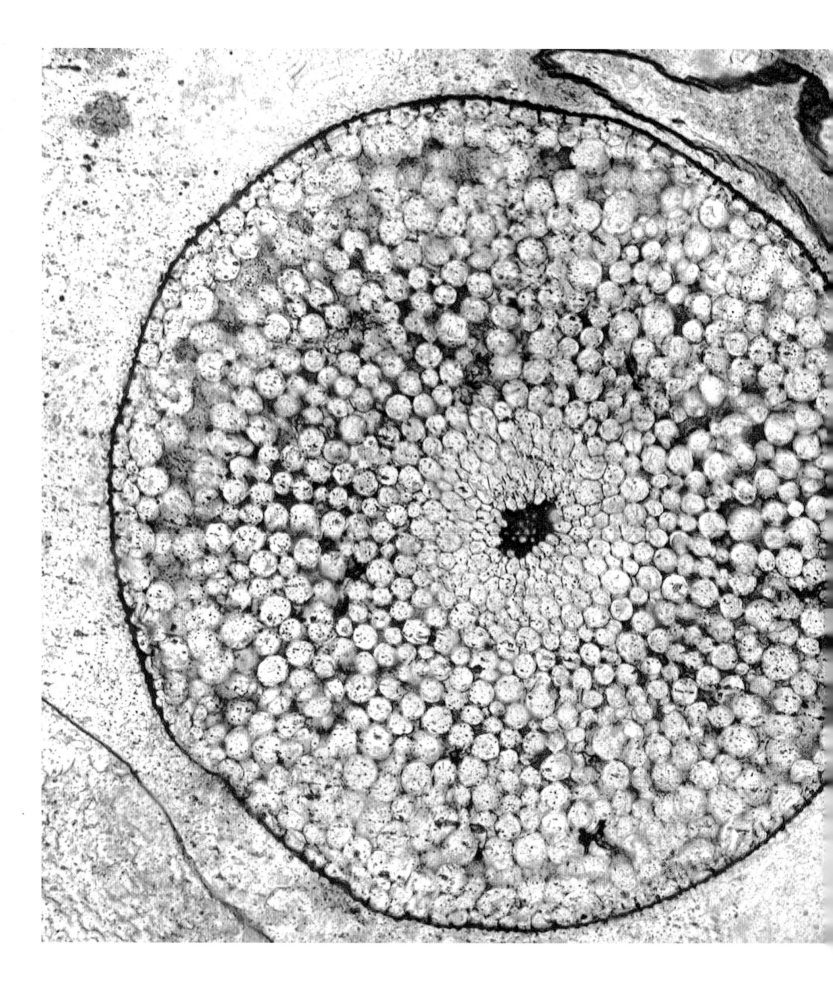

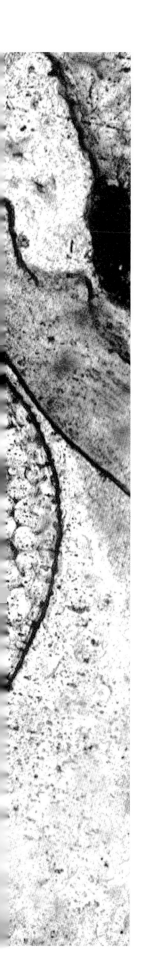

Pioneering vegetation

One of the earliest land plant pioneers, this fossilized *Thursophyton* had branching stems as long as a human hand. Ground-hugging plants like these provided shelter and food for the earliest land animals – paving the way for the first complex food chains on land.

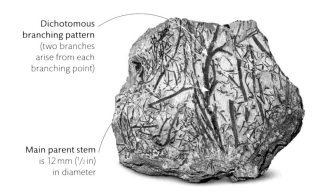

Dichotomous branching pattern (two branches arise from each branching point)

Main parent stem is 12 mm (½ in) in diameter

invading land

Land and air were inhospitable to early life, which evolved in water. As living bodies are at least 70 per cent water, any organism that ventures onto land must prevent water loss by evaporation. The first to do so – bacteria – probably colonized rock more than 3 billion years ago. When larger life forms, such as plants and animals, invaded land, they needed strengthening tissues to enable them to stand upright. The first land plants – descended from aquatic algae – appeared more than 400 million years ago and had waxy, waterproof shoots and leaves, stiff internal pipework, and roots providing anchorage.

Plant plumbing

This fossilized cross-section of an early mosslike plant, *Rhynia major*, shows a transport pipe – called a xylem vessel – running through the centre of a pithy stem. Just like in modern land plants, this pipe carried water from the roots and strengthened the upright stem.

MULTI-LAYERED COMMUNITY

The evolution of branching, erect land vegetation meant that plants, such as *Cooksonia*, could absorb more light energy for photosynthesis and disperse their reproductive spores on the wind. However, *Cooksonia* contributed only to the lowest level in new, multi-layered land communities. Taller pioneers called nematophytes (including *Germanophyton*, *Mosellophyton*, and *Prototaxites*) towered above them. Only known from the fossil record, they may have been related to non-photosynthetic fungi rather than plants.

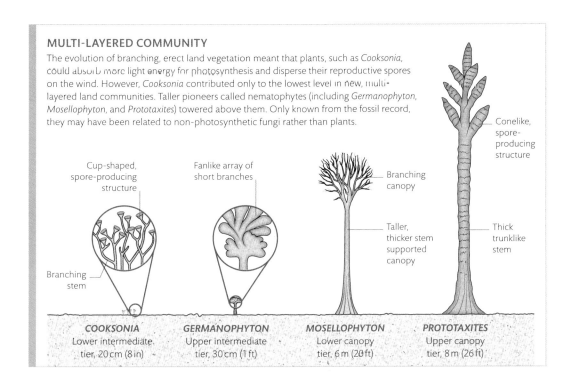

Cup-shaped, spore-producing structure

Branching stem

Fanlike array of short branches

Branching canopy

Taller, thicker stem supported canopy

Conelike, spore-producing structure

Thick trunklike stem

COOKSONIA	GERMANOPHYTON	MOSELLOPHYTON	PROTOTAXITES
Lower intermediate tier, 20 cm (8 in)	Upper intermediate tier, 30 cm (1 ft)	Lower canopy tier, 6 m (20 ft)	Upper canopy tier, 8 m (26 ft)

Primary succession in action
Anak Krakatau, a volcanic island off the
Indonesian islands of Java and Sumatra,
first appeared in 1927. Repeated eruptions
have limited the growth of vegetation, but
as seen in this satellite image, plant life is
gradually becoming established.

succession

The sequence of organisms that colonize newly available habitats is predictable, and
is known as succession. In areas where no life existed before, such as in the path
of lava flows or retreating glaciers, succession is considered primary, whereas after
disturbances that destroy vegetation but not soil, such as wildfires, succession is
secondary. In both types of succession, early colonizers (pioneer species) tend to
be fast-growing, rapidly reproducing species that capitalize on the wealth of new
nutrients and resources. Later arrivals are slower growing but hardier, ensuring their
survival. Early species may facilitate the arrival of later species by modifying the
environment, or they may inhibit invaders until they are outcompeted.

THE STAGES OF PRIMARY SUCCESSION
Primary succession occurs when completely new habitats arise, as on a cooled lava flow.
Barren rock is broken down by microorganisms, lichens, and algae, forming soil, which
supports rapidly growing plants. More complex ecosystems form as generations of
decaying plants build up the soil, allowing other plants to grow. Eventually, a relatively
stable and complex ecosystem may form. This is called the climax community.

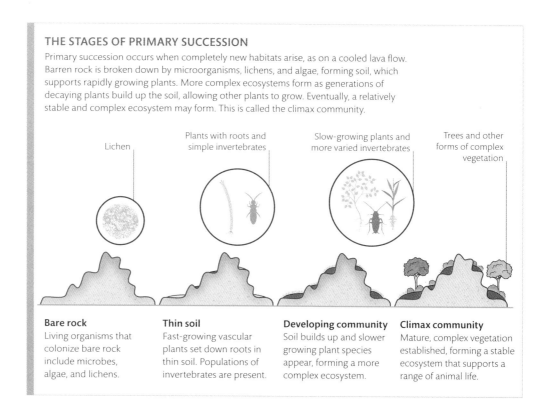

Lichen

Plants with roots and
simple invertebrates

Slow-growing plants and
more varied invertebrates

Trees and other
forms of complex
vegetation

Bare rock
Living organisms that
colonize bare rock
include microbes,
algae, and lichens.

Thin soil
Fast-growing vascular
plants set down roots in
thin soil. Populations of
invertebrates are present.

Developing community
Soil builds up and slower
growing plant species
appear, forming a more
complex ecosystem.

Climax community
Mature, complex vegetation
established, forming a stable
ecosystem that supports a
range of animal life.

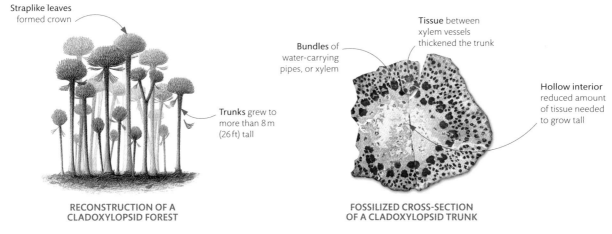

Straplike leaves formed crown

Trunks grew to more than 8 m (26 ft) tall

RECONSTRUCTION OF A CLADOXYLOPSID FOREST

Bundles of water-carrying pipes, or xylem

Tissue between xylem vessels thickened the trunk

Hollow interior reduced amount of tissue needed to grow tall

FOSSILIZED CROSS-SECTION OF A CLADOXYLOPSID TRUNK

Devonian forests
Fossils show that the first forests grew nearly 400 million years ago in the Devonian period. They were formed by distant cousins of ferns called cladoxylopsids, which were among the first plants to develop thick stems supported by rigid water-carrying pipes.

forests form

Plants that expose more of their foliage to the Sun's energy-giving rays generate, by their photosynthesis, more nourishment for growth. The first land plants sprawled on the ground – but as they competed for light, some evolved taller stems that allowed them to escape shadows cast by their neighbours. In the race to grow upwards to the light, they strengthened their stems with wood and evolved into the first trees. Today trees are the biggest single – non-colonial – organisms on the planet. And, growing together, they make up the most complex land habitats of all: forests, whose foliage can reach higher than a 15-storey building.

Today's rainforests
In the warm, moist tropics, conditions in the forest canopy stay humid and wet under regular rainfall, meaning that smaller plants – called epiphytes – can set root and grow on branches of trees. Here, beyond the shaded ground, a huge diversity of species – including bromeliads, orchids, and ferns – can thrive, providing a rich habitat for arboreal rainforest animals such as insects, spiders, frogs, and birds.

FOREST LAYERING

Trees cast shadows and create dark, humid microclimates close to the ground. Trees and shrubs, and the vines and epiphytes they carry, develop into layers, especially in rainforests where a huge number of species are adapted to different conditions. Undergrowth and understorey trees are most tolerant of shade and many even complete their life cycles there, whereas canopy and emergent trees need full exposure to the Sun to set seed.

LAYERS OF A RAINFOREST

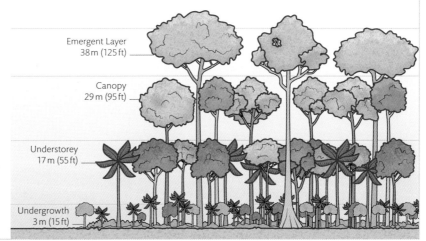

Emergent Layer 38 m (125 ft)

Canopy 29 m (95 ft)

Understorey 17 m (55 ft)

Undergrowth 3 m (15 ft)

fossilization

The process that transforms formerly living organisms into long-lasting remains, fossilization usually involves the replacement of original tissues with more durable minerals. This can happen in many different ways depending on the environment in which the animal or plant was buried, and the process can take millions of years to replace the minerals in a skeleton. The resulting remains can withstand the destructive forces of burial and erosion and retain their original shapes and features of the tissues, even down to the cellular structures, providing a wealth of information for today's scientists.

Fossil formation
The two main varieties of fossils are body fossils (parts of the life form itself) and trace fossils (evidence of its behaviour). Most body fossils, such as bones or teeth, have their tissues replaced with minerals, in a process called permineralization. Trace fossils preserve signs of a creature's activity, such as footprints or burrows.

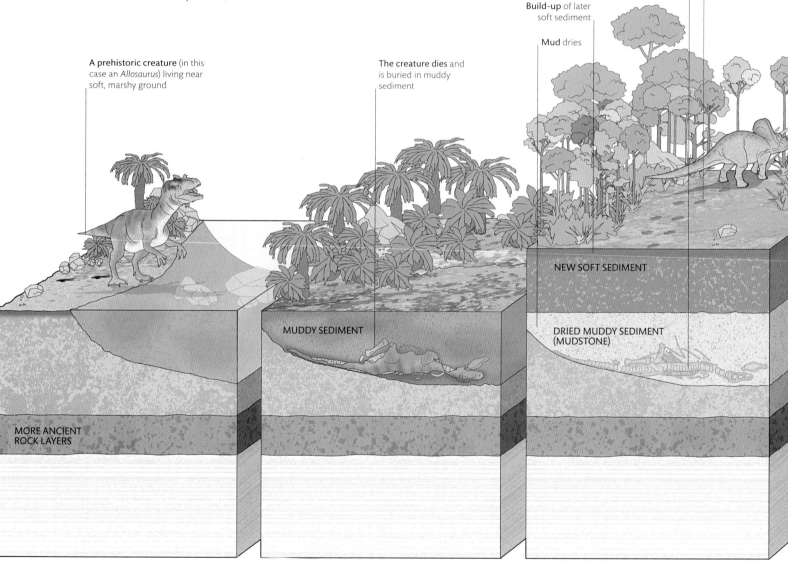

Footprints left in soft sediment by a later creature (*Triceratops*)

Minerals from surrounding sediments replace bone

Build-up of later soft sediment

Mud dries

A prehistoric creature (in this case an *Allosaurus*) living near soft, marshy ground

The creature dies and is buried in muddy sediment

NEW SOFT SEDIMENT

MUDDY SEDIMENT

DRIED MUDDY SEDIMENT (MUDSTONE)

MORE ANCIENT ROCK LAYERS

Ancient life
Creatures most likely to be preserved as fossils lived in places where there was a build-up of sediment in which they were buried after death.

Death and burial
If, after death, the creature's remains are protected from scavenging and further decay by overlying sediments for long periods of time, they may be fossilized.

Sediment build-up
Over time, soil and sediment build up and begin to turn to rock. During this process, the organic remains within the sediment are replaced by minerals.

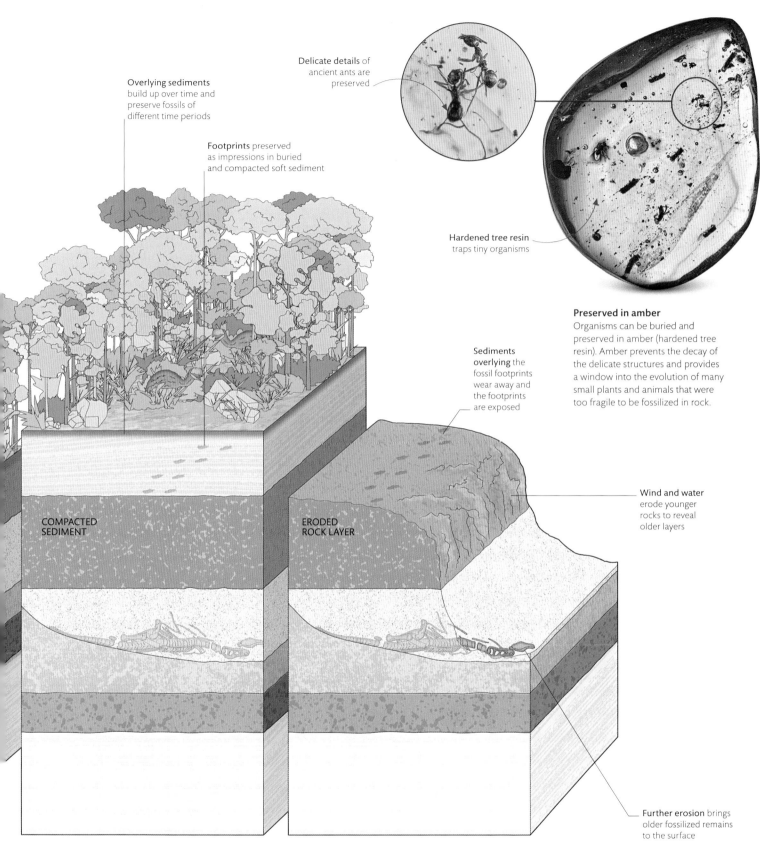

Overlying sediments build up over time and preserve fossils of different time periods

Footprints preserved as impressions in buried and compacted soft sediment

Delicate details of ancient ants are preserved

Hardened tree resin traps tiny organisms

Preserved in amber
Organisms can be buried and preserved in amber (hardened tree resin). Amber prevents the decay of the delicate structures and provides a window into the evolution of many small plants and animals that were too fragile to be fossilized in rock.

Sediments overlying the fossil footprints wear away and the footprints are exposed

Wind and water erode younger rocks to reveal older layers

COMPACTED SEDIMENT

ERODED ROCK LAYER

Further erosion brings older fossilized remains to the surface

Later fossils
Fossils of later life forms or fossil footprints may be preserved in newer rock layers. Footprints are formed as impressions or casts, depending on the type of rock.

Exposure and discovery
Fossils are usually found when they are exposed at the surface, in locations where erosion has removed overlying layers of sediment and rock.

coal forms

Black seams of coal are testament to the intimate link between the living biosphere and the rocks beneath it. All living things are part of a cycle of matter that persists long after they die and decay. Most are destined to rot completely into minerals that nourish new growth in other organisms. But under specific circumstances, decomposition slows almost to a halt. By 360 million years ago, plants had evolved a woody material called lignin that helped them grow into tall forests. However, decomposers such as fungi had not yet evolved the ability to break the lignin down. As a result, the woody trunks became compacted in the ground and – over many thousands of years – turned into carbon-rich coal.

Marks on trunk are scars left from falling leaves

Fossilized tree bark
Coal-forming swamp forests were composed mainly of spore-producing trees, such as this *Lepidodendron* – a relative of today's ferns.

Seam of coal

Coal is a type of sedimentary rock made almost entirely from black carbon, and most was deposited at least 100 million years before the time of the dinosaurs, during a period appropriately called the Carboniferous. Seams – like this one seen above ground along a beach in New South Wales, Australia – are the organic remains of Carboniferous swamp forests.

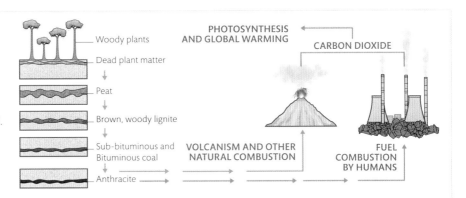

LONG CARBON CYCLE

Coal forms when dead vegetation becomes compacted. It is rotted down by bacteria into peat. This is buried and warmed under accumulated sediment, squeezing water and gases from the remains, producing first lignite and then coal. The process concentrates carbon that can only then be liberated by combustion into carbon dioxide (CO_2); faster, human-caused forms of combustion raise CO_2 levels.

Woody plants
Dead plant matter
Peat
Brown, woody lignite
Sub-bituminous and Bituminous coal
Anthracite

PHOTOSYNTHESIS AND GLOBAL WARMING

CARBON DIOXIDE

VOLCANISM AND OTHER NATURAL COMBUSTION

FUEL COMBUSTION BY HUMANS

oxygen-fuelled giants

Once microbes flooded the air with oxygen more than 2 billion years ago (see p.252), atmospheric levels of oxygen remained fairly steady. But during the Carboniferous, oxygen levels nearly doubled. That this happened at a time when coal seams were building up is no coincidence. As carbon was locked away underground, less of it combined chemically with oxygen, so oxygen gas built up in the air. Such a highly charged atmosphere meant more wildfires and the evolution of huge arthropods, such as millipedes the length of cars.

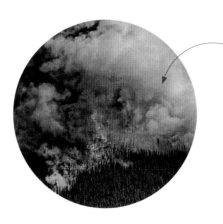

Fire consumes oxygen and emits ash, but also releases water vapour and carbon dioxide

Spontaneous forest fires
Prehistoric rocks show that forest blazes were common in the Carboniferous. Today, some plants found in hot, dry climates are fire resistant, while others actually need the fires for germination.

ATMOSPHERIC OXYGEN

Photosynthesis generates oxygen, but other processes consume it, such as fire, and respiration by living things. A balance between the two has helped to stabilize oxygen levels for much of prehistory. But a peak happened in the Carboniferous when woody trees evolved that initially resisted attack by decomposers. Less decomposer respiration meant that less oxygen was consumed, so levels rose.

PERIOD

CAMBRIAN ORDOVICIAN SILURIAN DEVONIAN CARBONIFEROUS PERMIAN TRIASSIC JURASSIC CRETACEOUS PALEOGENE NEOGENE

ATMOSPHERIC OXYGEN PERCENTAGE

45
40
35
30
25 — CURRENT OXYGEN LEVEL
20
15
10
5
0

539 485 444 419 359 299 252 201 145 66 23

AGE/MYA

CHANGING OXYGEN LEVEL

Oxygen level rose to 35%

Oxygen levels are 21%

Largest flying insect
Insects rely on oxygen seeping directly into their tissues through perforations in their body wall. The high oxygen levels of the Carboniferous could diffuse deeper into tissues, so could satisfy the demands of bigger bodies. The result was giants like this dragonfly-like insect, *Meganeura*, which had a 70-cm (28-in) wingspan.

walking on land

Once algae and plants had colonized the land from the oceans where life had begun, animals followed – the vegetation providing shelter and food. The first animals to venture out of water probably did so fleetingly, perhaps at night, just as some marine snails do today. Prehistoric footprints indicate that shrimplike animals made the move from water to land more than half a billion years ago. Their armour-plated exoskeleton doubtless helped shield them from the Sun's drying rays. But it took millions of years of evolution to turn the tentative pioneers into permanent air-breathing insects and spiders. By the time of the coal-producing swamp forests (see pp.270–71), backboned animals (vertebrates) had shifted landwards, and a fleshy finned fish had evolved into primitive walking tetrapods.

Walking vertebrate
Living between 295 and 272 million years ago, *Seymouria*, an animal measuring 60 cm (2 ft) in length that resembled a giant modern day salamander, had limbs strong enough to lift its body off the ground. Like all land-living backboned animals, it evolved from a group of fish whose thick fins had bony supports that would become the digits of walking feet.

Pectoral fins are used as forelimbs to clamber about on land

Reinvasion of the land
Today many groups of fish continue to evolve ways of conquering the land. Descended from oceanic gobies, gold-spotted mudskippers (*Periophthalmus chrysospilos*) have forward-positioned pectoral (chest) fins that are strong enough to haul them across mudflats.

MULTIPLE LAND INVASIONS

During the history of their evolution, animals have conquered the land several times – in independent groups. By comparing DNA of living animals to date their ancestry, and studying fossil records, scientists estimate when different groups of land animals developed from their aquatic ancestors. The oldest land animals – jointed-legged arthropods (including insects, crustaceans, and spiders), and nematode worms – had already filled the prehistoric forests by the time the first four-legged vertebrates, or tetrapods, walked on land. All of these pioneers started out being partly aquatic, but eventually evolved the means to stay out of water. Snails, slugs, and earthworms turned landward much later – around the time of the dinosaurs.

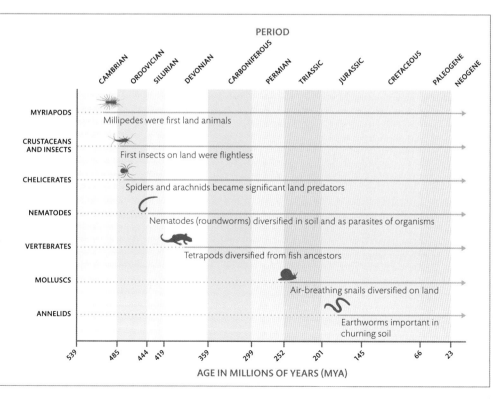

PERIOD

CAMBRIAN · ORDOVICIAN · SILURIAN · DEVONIAN · CARBONIFEROUS · PERMIAN · TRIASSIC · JURASSIC · CRETACEOUS · PALEOGENE · NEOGENE

MYRIAPODS
Millipedes were first land animals

CRUSTACEANS AND INSECTS
First insects on land were flightless

CHELICERATES
Spiders and arachnids became significant land predators

NEMATODES
Nematodes (roundworms) diversified in soil and as parasites of organisms

VERTEBRATES
Tetrapods diversified from fish ancestors

MOLLUSCS
Air-breathing snails diversified on land

ANNELIDS
Earthworms important in churning soil

539 · 485 · 444 · 419 · 359 · 299 · 252 · 201 · 145 · 66 · 23

AGE IN MILLIONS OF YEARS (MYA)

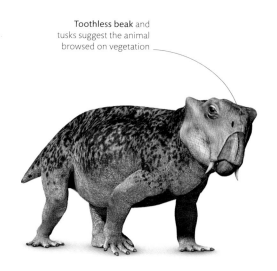

Toothless beak and tusks suggest the animal browsed on vegetation

Evidence of extinction
The dark layers visible in Siberia's Putorana Mountains built up as lava flooded the land from volcanic fissure eruptions, in which Earth's crust fractured over a wide area. The lava solidified in layers to form a vast plateau, of which this is a small part. This enormous volcanic event came as close to eliminating complex life as at any time in Earth's history.

The Permian's great survivor
A distant reptilian predecessor of mammals, *Lystrosaurus* accounts for up to 95 per cent of vertebrate fossils deposited after this event, and a burrowing habit may have protected it.

the great dying

Earth has been described as the "Goldilocks planet" – for most of its existence, conditions have been just right for living things to thrive and evolve. But several events have caused colossal loss of species, or mass extinctions. One of these had an extra-terrestrial cause – an asteroid strike that killed the dinosaurs (see p.295) – but most events were triggered by Earth's own violent geology. The biggest happened 250 million years ago at the end of the Permian – before the dinosaurs. The climate change that resulted from huge volcanic eruptions wiped out more than three-quarters of all species. But there are always winners as well as losers, and from the survivors evolved new kinds of organisms that took over the biosphere.

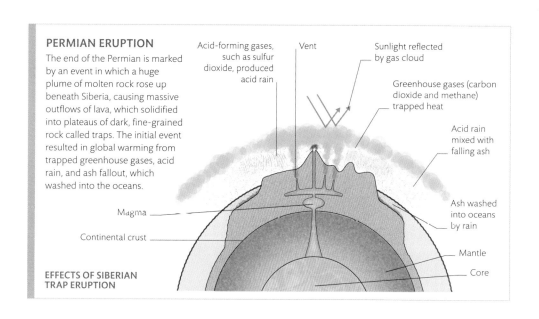

PERMIAN ERUPTION
The end of the Permian is marked by an event in which a huge plume of molten rock rose up beneath Siberia, causing massive outflows of lava, which solidified into plateaus of dark, fine-grained rock called traps. The initial event resulted in global warming from trapped greenhouse gases, acid rain, and ash fallout, which washed into the oceans.

Acid-forming gases, such as sulfur dioxide, produced acid rain

Vent

Sunlight reflected by gas cloud

Greenhouse gases (carbon dioxide and methane) trapped heat

Acid rain mixed with falling ash

Ash washed into oceans by rain

Magma

Continental crust

Mantle

Core

EFFECTS OF SIBERIAN TRAP ERUPTION

Blind trilobite had eyeless cephalon

CONOCORYPHE

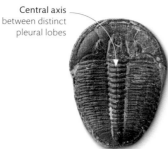

Central axis between distinct pleural lobes

ELRATHIA

Long, crescent-shaped eyes

OLENELLUS

Body was up to 50 cm (20 in) long, from head to tail

PARADOXIDES

Rise of the trilobites

Trilobites originated early in the Paleozoic era as part of the Cambrian explosion 539–520 million years ago (see pp.256–57), and their durable exoskeletons form a major component of its fossil record. Trilobites were very successful and reached a peak in their diversity towards the end of the Cambrian (485 million years ago), and into the Ordovician.

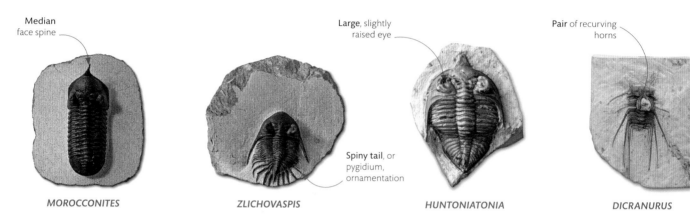

Median face spine

MOROCCONITES

Spiny tail, or pygidium, ornamentation

ZLICHOVASPIS

Large, slightly raised eye

HUNTONIATONIA

Pair of recurving horns

DICRANURUS

Decline of the trilobites

Throughout the Silurian and Devonian periods that followed the late Ordovician mass extinction, the number of trilobite families continued to decrease, although many groups were still lavishly ornamented. Only one order of trilobites, the Proetida, survived past the Devonian, and finally went extinct in the Permian mass extinction (see pp.276–77).

trilobite diversity

Trilobites were marine arthropods (animals with an exoskeleton, segmented body, and paired, jointed appendages), part of the group that includes today's crustaceans, arachnids, and insects. The trilobite body was divided into three segments: a cephalon (head), thorax, and pygidium (tail). The mouth opened on the underside, below the cephalon. However, the name trilobite, meaning "three-lobed", is due to the thorax being divided into three parts – a central axial lobe, and a pleural lobe on each side, each pleural lobe bearing legs with gills. With this body plan, trilobites diversified in the oceans and lived for 270 million years, leaving a rich fossil record of more than 20,000 species.

Three-pronged tridentlike horn extended from front of cephalon

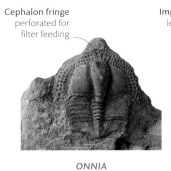

Cephalon fringe
perforated for
filter feeding

ONNIA

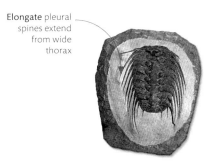

Impressions of
legs visible in
the rock

TRIARTHRUS

Elongate pleural
spines extend
from wide
thorax

SELENOPELTIS

Ornamented central
section of cephalon,
called the glabella

ENCRINURUS

Turnover of groups

During the Ordovician period new groups of trilobites with distinctive body plans continued to develop, but the number of families steadily decreased. During the late Ordovician mass extinction, about 445 million years ago, many of the groups originating in the Cambrian went extinct, but were replaced by those that had evolved in the Ordovician.

Mobile, jointed
thorax segments
allowed for
flexibilty

Trident trilobite

Although trilobites declined during the Devonian, some, such as this *Walliserops*, were magnificent. The function of its distinctive cephalon horn is unclear – researchers believe it may have served in sexual display, for battling rivals, or as a signal of a healthy individual – all functions served by deer antlers today.

Defensive pleural
spines

Tall recurved brow
horn above eye

Complex
multilensed eye

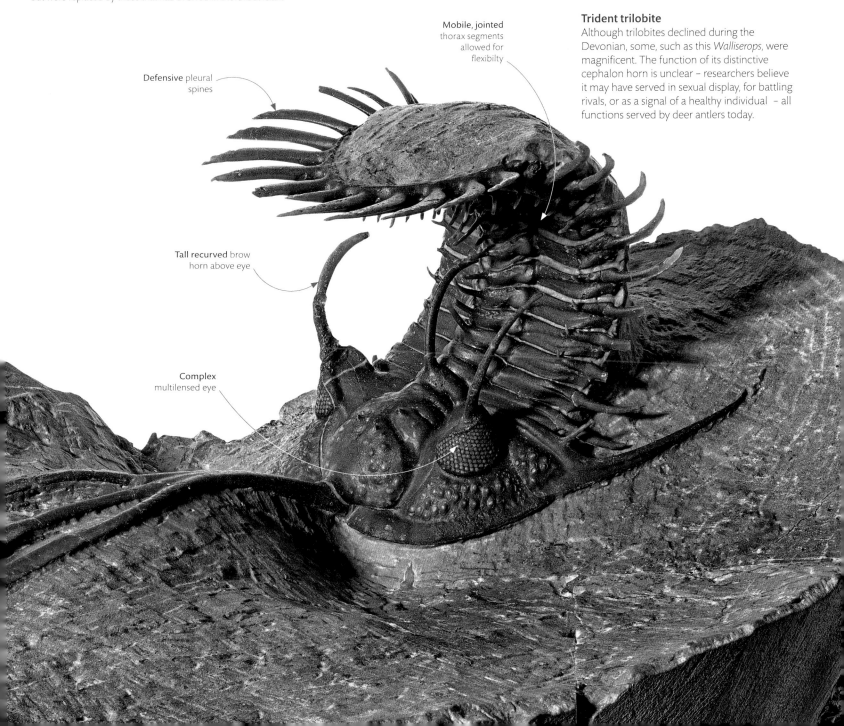

surviving drought

All organisms need water, but many have evolved ways of living in the driest habitats. As continents shift positions, land turns to desert for different reasons and places can be so far inland that rainfall scarcely reaches them. Around 250 million years ago – just before the rise of the dinosaurs – most of Earth's land was amassed into a supercontinent called Pangaea with a vast desert interior, and global temperatures peaked. Amphibians that laid eggs without shells in water, declined. But reptiles, whose hard-shelled eggs developed on land, could thrive in the driest regions.

Dry interior of Mongolia
In the centre of continental Asia, the Gobi Desert is in a "rain shadow" created by the Tibetan Plateau, which prevents moisture-laden winds from reaching it. With its extremes of temperature – bitterly cold winters and hot, dry summers – vegetation is sparse and only specialist desert animals survive.

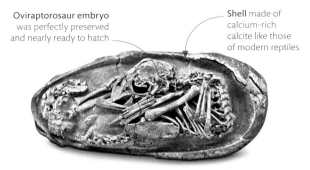

Oviraptorosaur embryo was perfectly preserved and nearly ready to hatch

Shell made of calcium-rich calcite like those of modern reptiles

Hard-shelled egg
The evolution of shelled eggs – such as this 66-million-year-old fossil – was critical in allowing reptiles to breed independently of water bodies.

ADAPTING TO ARIDITY

The aridity, or dryness, of a habitat is determined by the relative levels of rainfall and evaporation. A subhumid desert loses no more than twice the volume of its annual rainfall. But the driest desert can evaporate 200 times more water than it receives, so living things require extreme adaptations to extract it from their surroundings.

KEY

→ Water evaporation
◊ Rainfall

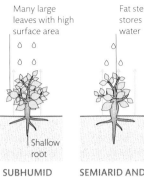

Many large leaves with high surface area

Shallow root

SUBHUMID DESERT

Fat stem stores water

SEMIARID AND ARID DESERT

Small fleshy leaves

Very fat (succulent) stem

Deep root

HYPERARID DESERT

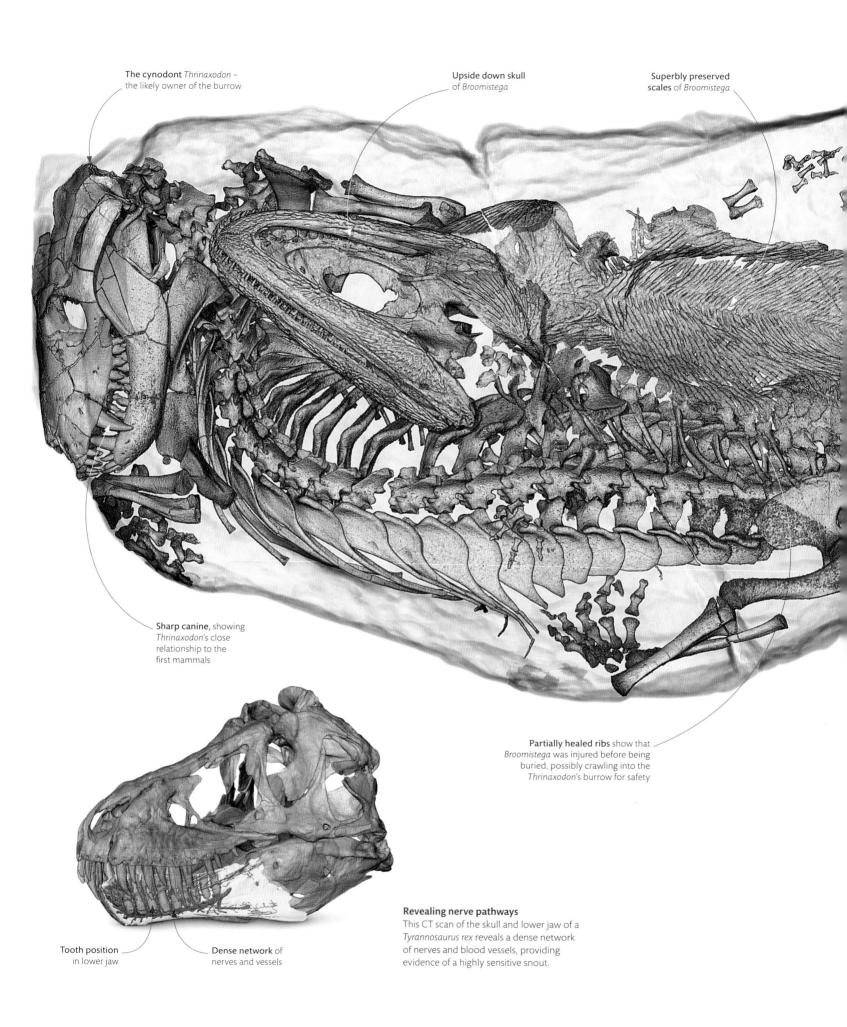

The cynodont *Thrinaxodon* – the likely owner of the burrow

Upside down skull of *Broomistega*

Superbly preserved scales of *Broomistega*

Sharp canine, showing *Thrinaxodon*'s close relationship to the first mammals

Partially healed ribs show that *Broomistega* was injured before being buried, possibly crawling into the *Thrinaxodon*'s burrow for safety

Tooth position in lower jaw

Dense network of nerves and vessels

Revealing nerve pathways
This CT scan of the skull and lower jaw of a *Tyrannosaurus rex* reveals a dense network of nerves and blood vessels, providing evidence of a highly sensitive snout.

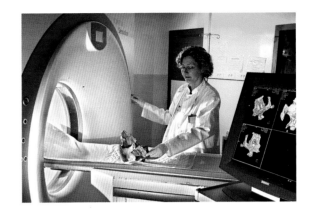

Looking inside a dinosaur
The internal structures of the fossilized skull of an *Arcovenator*, a theropod dinosaur from France, are viewed in great detail using a CT scanner.

history of Earth science

scanning fossils

The external features of bones and teeth reveal only some aspects of the biology of long-extinct animals. To investigate other characteristics, scientists must look inside the skeleton, where soft tissues once existed. In the past, this could only be done by destroying parts of the fossil. Modern scanning methods avoid this problem, and also enable us to investigate fossils that cannot be removed from the rock in which they are embedded.

Since the 1970s, X-ray technologies have been used by scientists to look inside fossils. High-powered X-rays penetrate the rock containing the fossil and show the different densities of the internal structures. The advent of more powerful and accessible computing in the 1980s and 1990s led to wider use of CT (computerized tomography) scanning by palaeontologists. In this technique, hundreds of X-ray images from different angles are aligned by the computer, creating a three-dimensional rendition of the interior of the fossil.

CT scanning techniques have steadily improved, allowing scientists to digitally reconstruct the internal anatomy of

ancient organisms with unprecedented detail. Researchers have now visualized the brains of long-extinct animals, including dinosaurs, marine reptiles, and early mammals. By comparing the sizes of different parts of the brain, digital models can be used to understand the importance to the creature of senses such as eyesight or smell.

CT scanning also shows other structures, such as the delicate chambers of the inner ear, nerve networks, and even daily growth layers in teeth. Today, new and more powerful imaging tools – for example, the synchrotron particle accelerator – reveal ever-more detail, even down to microscopic levels.

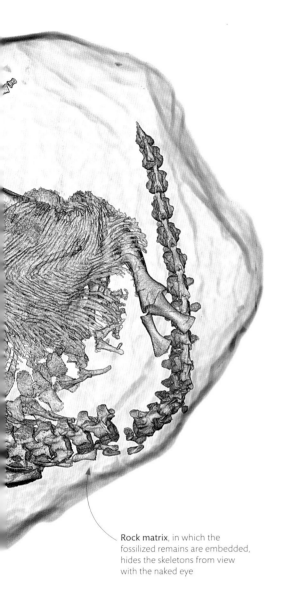

Rock matrix, in which the fossilized remains are embedded, hides the skeletons from view with the naked eye

Strange bedfellows
Thrinaxodon (greyish pink), a type of extinct mammal-like animal called a cynodont, and a *Broomistega* (dark grey), an amphibian, were buried together in a Triassic burrow, where the *Thrinaxodon* may have been lying dormant to avoid a summer drought. This amazing double fossil was revealed only when scientists used a high-powered synchrotron particle accelerator to scan rock that they suspected of containing a fossilized cynodont burrow.

Almost all the traditional problems associated with recovering fossil data from rocks can be overcome with modern 3D imaging.

JOHN CUNNINGHAM *ET AL. A VIRTUAL WORLD OF PALEONTOLOGY* 2014

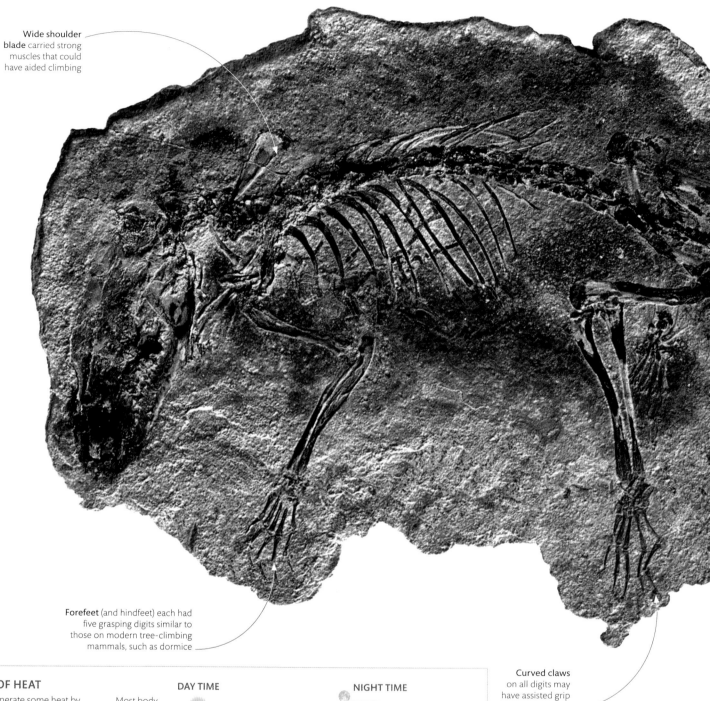

Wide shoulder blade carried strong muscles that could have aided climbing

Forefeet (and hindfeet) each had five grasping digits similar to those on modern tree-climbing mammals, such as dormice

Curved claws on all digits may have assisted grip when climbing

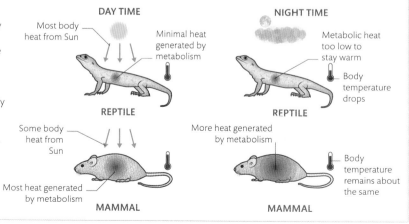

SOURCES OF HEAT

All animals generate some heat by their metabolism. Cold-blooded (endothermic) reptiles rely on the Sun for heat, but warm-blooded (ectothermic) animals, including mammals, can produce so much heat that they can keep their body temperature high, even when surroundings are cold. They do it by burning specialized brown fat. Moreover, they can also control their body temperature day and night via a gland in their brain called the hypothalamus.

DAY TIME

Most body heat from Sun

Minimal heat generated by metabolism

REPTILE

NIGHT TIME

Metabolic heat too low to stay warm

Body temperature drops

REPTILE

Some body heat from Sun

Most heat generated by metabolism

MAMMAL

More heat generated by metabolism

Body temperature remains about the same

MAMMAL

Furry fossil

The rock around the fossilized bones of this shrew-sized *Eomaia*, meaning "dawn mother", show the impression of a coat of fur: clear evidence that this was a mammal. *Eomaia* lived 125 million years ago in China – during the Cretaceous peak of the dinosaurs – and was an ancient "sister" of living marsupial and placental mammals (see pp.298–99), but too early to be placed in either of these groups.

Hedgehogs inhabit grasslands, hedgerows, and woodland, searching for insects, beetles, and worms

Predatory mammal

Most living mammals, like this European hedgehog (*Erinaceus europaeus*), are nocturnal – suggesting that night foraging was adopted early in the group's evolution. Day-time habits emerged in later groups, such as squirrels and monkeys.

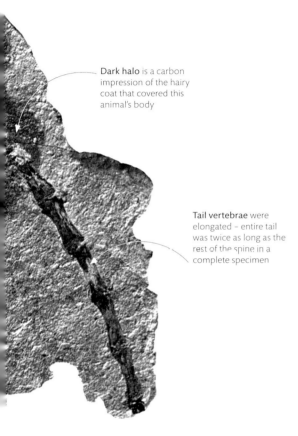

Dark halo is a carbon impression of the hairy coat that covered this animal's body

Tail vertebrae were elongated – entire tail was twice as long as the rest of the spine in a complete specimen

becoming
warm-blooded

While cold-blooded reptiles walked the land and were warmed by the Sun's rays, some of their descendants found different ways of surviving. These were the mammals, or warm-blooded animals. They adopted a more frenetic life: their sharp-toothed fossils point to them chasing scurrying insect prey, and their skeletons show evidence that they matured quickly. Some fossils even have impressions of a novel innovation: hairy skin. Their diet, rich in fat and protein, fuelled a body that stayed warm, even during cold nights, which helped them avoid competition with reptiles that were mainly active by day. Insulated by a furry coat, their descendants could survive in permanently cold habitats, such as Arctic tundra, that was inhospitable to their cold-blooded cousins.

Reconstruction of *Eomaia*

Long, clawed digits and a long tail that could have helped with balance suggest that *Eomaia* was an agile small animal that might have climbed trees.

Thick furry coat
helped trap body heat

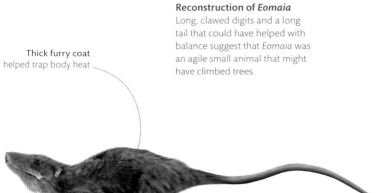

The Red Deer River Valley, incised like a wound into the prairies of Alberta, Canada, is home to some of the richest dinosaur fossil beds in the world. The striped hills provide a record of life in the late Cretaceous (78–66 million years ago) along the shores of an ancient intercontinental sea. The valley was carved by the drainage of glacial lakes after the last ice age (11,700 years ago), which exposed the soft Cretaceous sediments.

Red Deer River Valley

This valley is home to numerous First Nations peoples, who were the first fossil collectors here. An area called the badlands produces the *iniskim* – buffalo-shaped stones sacred to the Niitsitapi (Blackfoot), formed from ammonite fossils.

In the 1880s geological surveys recognized the incredibly rich geological resources of the valley, including thick coal seams, as well as an abundance of dinosaur fossils. Some areas in the valley, like Dinosaur Provincial Park or the Horsethief Canyon, are so rich in fossils that palaeontologists walk over some fossil skeletons to reach the most scientifically important examples. The fossils found here reveal some of the best known dinosaur ecosystems, spanning 12 million years in time, and include familiar groups like the iconic tyrannosaurs, duck-billed hadrosaurs, frilled ceratopsians, and fearsome dromaeosaurs. Living alongside these dinosaurs in the lush subtropical climate was a diverse array of fish, lizards, turtles, crocodiles, and early mammals.

Fossil is curled into the "death pose", common among dinosaur skeletons found in the valley

Complete Gorgosaurus

Painted valley
The colours of the hills in the Horsethief Canyon are caused by variations in rock type and reflect different environments. The mudstones (brown), sandstones (white), ironstones (orange), and coal seams (black) are home to millions of fossils, giving palaeontologists a clear picture of the ecosystems of the Late Cretaceous.

THE ORIGINS OF POWERED FLIGHT

The first animals that took to the skies where insects. Their wings evolved as flaplike extensions of their exoskeleton. Backboned fliers adapted their forelimbs as wings. Birds then used a fringe of stiff flight feathers as their wing blade. Other vertebrates used a stretch of skin instead – between the fingers of bats, or between finger and ankle of pterosaurs.

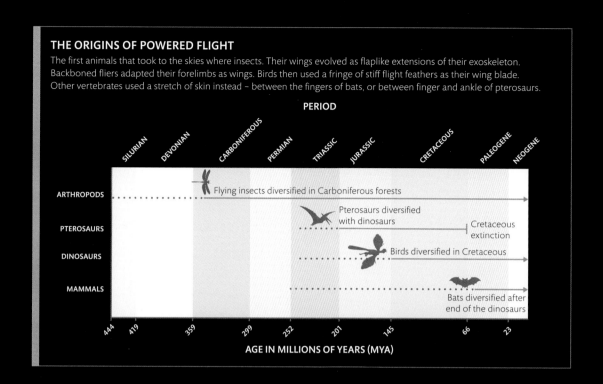

PERIOD

SILURIAN · DEVONIAN · CARBONIFEROUS · PERMIAN · TRIASSIC · JURASSIC · CRETACEOUS · PALEOGENE · NEOGENE

ARTHROPODS — Flying insects diversified in Carboniferous forests

PTEROSAURS — Pterosaurs diversified with dinosaurs — Cretaceous extinction

DINOSAURS — Birds diversified in Cretaceous

MAMMALS — Bats diversified after end of the dinosaurs

444 419 359 299 252 201 145 66 23

AGE IN MILLIONS OF YEARS (MYA)

taking to the air

Many living things break free from gravity and rise into the atmosphere, which becomes an aerial highway for crossing distances quickly, largely unhindered by friction, as well as a potential source of airborne food. Plants and fungi scatter seeds, pollen, and spores, while flying animals evolved wings – aerofoils that are sufficiently broad to generate lift, but thin enough to keep them lightweight. Many animals – such as "flying" squirrels – use static wings just to glide (see p.299). But other animals use flapping wings to generate forward thrust – which allows them to exploit the air to a much greater extent; a few birds today can stay airborne for months at a time.

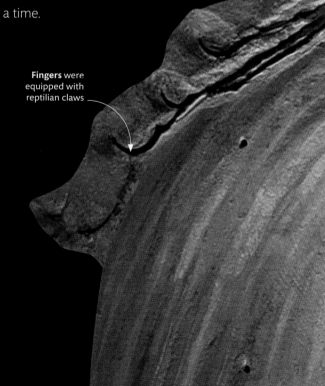

Fingers were equipped with reptilian claws

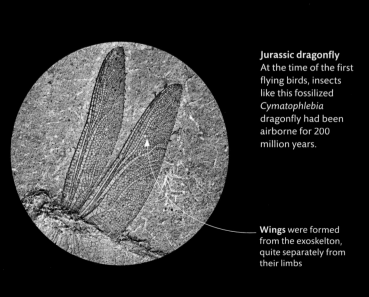

Jurassic dragonfly
At the time of the first flying birds, insects like this fossilized *Cymatophlebia* dragonfly had been airborne for 200 million years.

Wings were formed from the exoskelton, quite separately from their limbs

Earliest birdlike animal
All flying animals evolved from land-bound ancestors. Birds descended from two-legged predatory dinosaurs. But *Archaeopteryx*, one of the oldest flying dinosaurs, dating back 150 million years, shows transitional features. It had reptilian teeth and tail, but avian, or birdlike, beak and feathers. Scientists are not certain if it flapped its wings or was a glider.

Impressions of long, stiff flight feathers visible; the wings acted as aerofoils creating lift in flight

Small, sharp teeth suggest a diet of small animals, such as insects

Birdlike pointed beak

Tail feathers were supported by a long, bony reptilian tail; in modern birds they are rooted on a bony stump

giants of the land

Dinosaurs lived in the Mesozoic era, 252–66 million years ago, and included the largest animals that ever lived on land. Animals grow big by converting the food they eat into flesh, and giants have huge appetites; dinosaurs probably managed this by combining appetite with efficient physiology. Their breathing relied on a system of supplementary air sacs that helped their lungs extract more oxygen from air. In today's birds – descendants of the dinosaurs (see p.288) – air sacs lighten the body, which helps with flight. In dinosaurs, the air sacs eased muscular effort, which helped with walking. An oxygen-enriched fast metabolism also meant speedy growth: estimates suggest that these giants could have gained 2 tonnes in body weight each year.

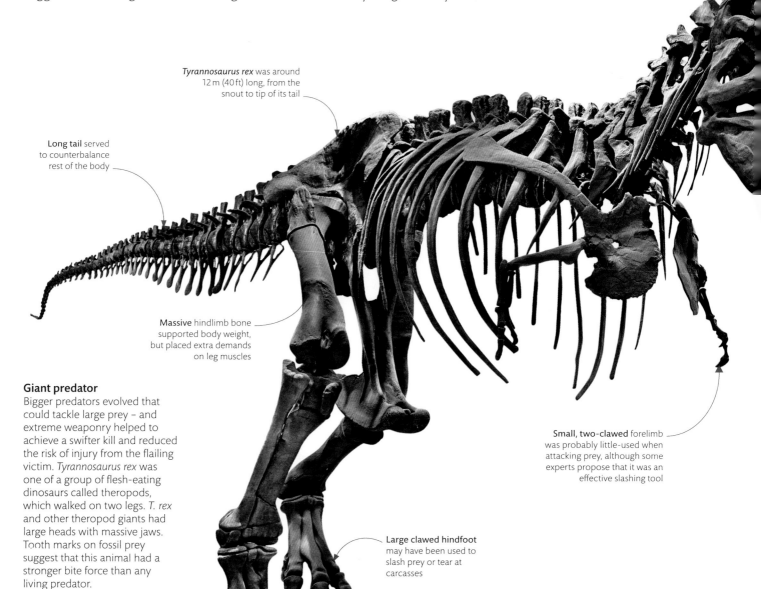

Small jaws had small, peglike teeth that could crop leaves, but not chew them

Tyrannosaurus rex was around 12 m (40 ft) long, from the snout to tip of its tail

Long tail served to counterbalance rest of the body

Massive hindlimb bone supported body weight, but placed extra demands on leg muscles

Small, two-clawed forelimb was probably little-used when attacking prey, although some experts propose that it was an effective slashing tool

Large clawed hindfoot may have been used to slash prey or tear at carcasses

Giant predator

Bigger predators evolved that could tackle large prey – and extreme weaponry helped to achieve a swifter kill and reduced the risk of injury from the flailing victim. *Tyrannosaurus rex* was one of a group of flesh-eating dinosaurs called theropods, which walked on two legs. *T. rex* and other theropod giants had large heads with massive jaws. Tooth marks on fossil prey suggest that this animal had a stronger bite force than any living predator.

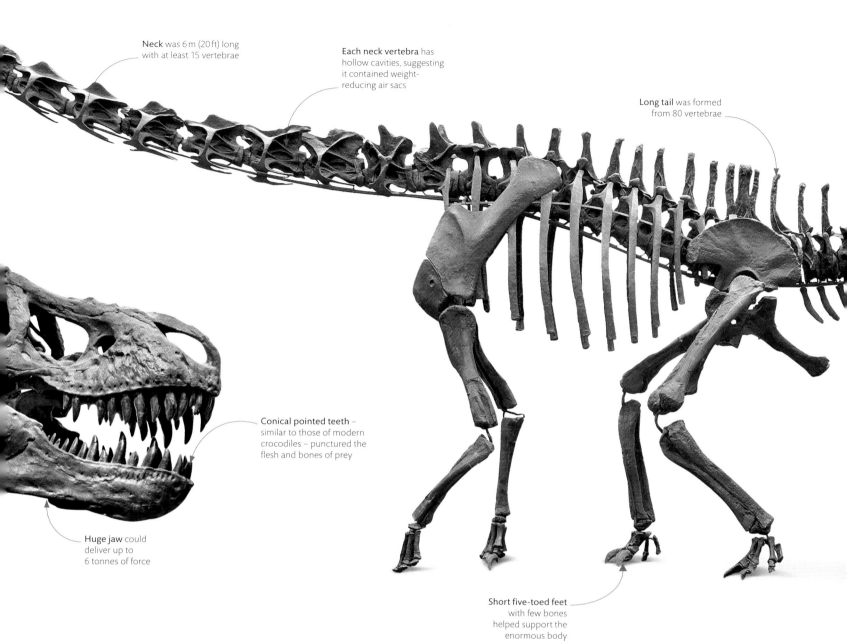

Neck was 6 m (20 ft) long with at least 15 vertebrae

Each neck vertebra has hollow cavities, suggesting it contained weight-reducing air sacs

Long tail was formed from 80 vertebrae

Conical pointed teeth – similar to those of modern crocodiles – punctured the flesh and bones of prey

Huge jaw could deliver up to 6 tonnes of force

Short five-toed feet with few bones helped support the enormous body

LIFESTYLE FROM FOOTPRINTS

Fossilized trackways can be used to estimate the walking speed of dinosaurs. By measuring the distance between successive prints of forefeet and hindfeet, scientists can estimate the length of an animal's stride. When this is combined with the shoulder-to-hip length, known from excavated skeletons, scientists can also calculate the dinosaur's likely speed. The tracks of a large sauropod called *Titanopodus* suggest a slow, ponderous gait – less than 5 km/h (3 mph) – which is consistent with that of a large herbivore that spent a long time digesting food.

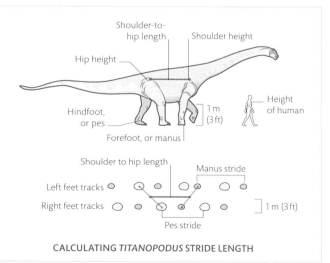

Shoulder-to-hip length

Shoulder height

Hip height

Height of human

1 m (3 ft)

Hindfoot, or pes

Forefoot, or manus

Shoulder to hip length

Manus stride

Left feet tracks

Right feet tracks

1 m (3 ft)

Pes stride

CALCULATING *TITANOPODUS* STRIDE LENGTH

Giant herbivore

The biggest dinosaurs of all were the herbivorous longnecked sauropods, such as *Diplodocus*, which averaged 27 m (90 ft) in length. Their teeth indicate that they swallowed mouthfuls of leaves whole, then relied on an enormous gut to digest food over a long time. Their long neck could sweep their head through a wide arc, enabling them to browse a wide area while remaining in one place.

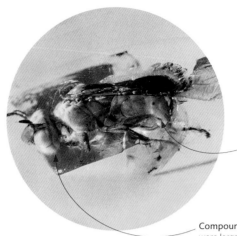

Fossilized bee

The groups of insects that rose to prominence after the demise of the dinosaurs included bees that diversified alongside flowering plants. This bee, from 50 million years ago, is embedded in Baltic amber – the solidified sticky resin of trees that trapped many small prehistoric animals, including flying insects.

Hair-fringed cavity
on hind leg was used to carry pollen

Compound eyes
were larger and likely had more colour receptors than those of beetles

Antenna has 11 segments and may have carried sensors for detecting pungent cycad cones, like those of today's boganiids

THE EARLY POLLINATORS

The Cretaceous (145–66 million years ago) is best known as the final period of the dinosaurs' reign. But it was also a time of major change in vegetation and insects: cone-bearing seed plants, such as cycads, declined as flowering plants diversified, and with that, their pollinators changed too. The cone-bearers had relied on beetles and flies, and the first flowers were probably pollinated by them too, but the showy, sweet-scented flowers later attracted hymenopteran insects, such as bees, and supplied nectar for their specially adapted lapping mouthparts.

CRETACEOUS POLLINATORS

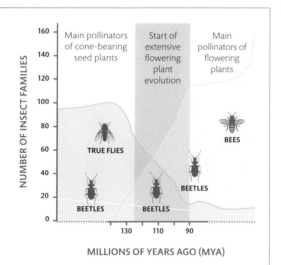

NUMBER OF INSECT FAMILIES

Main pollinators of cone-bearing seed plants

Start of extensive flowering plant evolution

Main pollinators of flowering plants

TRUE FLIES

BEETLES

BEETLES

BEETLES

BEES

160
140
120
100
80
60
40
20
0

130   110   90

MILLIONS OF YEARS AGO (MYA)

THE FIRST FLOWERS

By comparing flowers from a range of living plant families with fossils, biologists have reconstructed the most likely appearance of an early Cretaceous flower. The robust, thickly petalled bloom was similar to a magnolia. Like magnolias – and the cycads that dominated before – it was probably pollinated by beetles with poor colour vision and chewing mouthparts. Pollen, nectar, and parts of the flower may also have been food sources for the insects.

RECONSTRUCTION OF AN ANCESTRAL FLOWER

Middle whorls of stamens (male parts) produced pollen that brushed onto beetles

Inner spiral of carpels (female parts) contained eggs fertilized by pollen from other plants

Outer whorls of thick petals resisted chewing mouthparts of pollinating beetles

Cycad pollinator

This *Cretoparacucujus* beetle from around 100 million years ago is so well-preserved in fossilized amber that it can be confidently classified in the family of boganiid beetles, which still pollinate cycads today. And, tellingly, cycad pollen grains were found in the same piece of amber. Cycads produce a large amount of pollen – some is sacrificed to feed the beetles, and some is carried by beetles to fertilize the plant's neighbours.

insect pollinators

Plants have had varying fortunes with the animal life around them. Herbivores drove many plants to make unpalatable chemicals, while animals responded by evolving ways to detoxify them. But during the time of the dinosaurs, insects evolved a mutually beneficial relationship with seed plants: insects carried pollen for the plant's sexual reproduction, while plants rewarded the carriers with a meal. It began with cone-bearing cycads – distant cousins of wind-pollinated conifers (see p.309) – but the arrangement was so successful that it changed the face of the living planet. A new kind of seed plant evolved, with flowers instead of cones, and Earth blossomed.

Hair-lined cavity near base of mandible (jaw) collected pollen grains, which were transferred to the next cycad

Elongated clypeus, the basal part of the upper lip, one of the features that indicates this animal belongs to the boganiid family of beetles

Small compound eyes probably lacked the complex colour vision of the later flower-pollinating bees

Gently arched
spinal column

Distinctive "frill"
was formed of
solid bone

Strong limbs were
needed to support
and move animal's
massive body

Fosilized spiral shell
once enclosed a soft
tentacled body

Lost molluscs
The end-of-Cretaceous
extinction was as devastating
for marine life as it was to land
species – and many groups
that lived alongside the
dinosaurs disappeared too.
Among them were ammonites,
the spiral-shelled relatives of
modern squid and octopus.

The last dinosaurs

By the end of the Cretaceous period Earth was rich with dinosaurs, such as the three-horned *Triceratops*. But their long successful reign was already being threatened by massive volcanic eruptions in Asia – the so-called Deccan Traps – that spewed toxic fumes. The asteroid impact was their coup de grâce: an abrupt break in the fossil record suggests that within millennia, it had killed them all. *Triceratops* had an array of effective defensive and feeding adaptations, but these did not prepare it for this catastrophe.

ASTEROID EVIDENCE

In 1978, geophysicists uncovered anomalies in rocks that suggested the Gulf of Mexico had been struck by a massive asteroid. The impact left a crater 200 km (125 mile) in width centred near Chicxulub Pueblo at the northern tip of what is now the Yucatán peninsula. A thin band of iridium, an element common in asteroids, was then discovered in end-of-Cretaceous rocks all around the world – showing how the smashed remains of the asteroid were scattered far and wide, having a truly global effect.

THE CHICXULUB CRATER

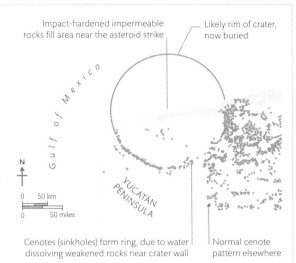

Impact-hardened impermeable rocks fill area near the asteroid strike

Likely rim of crater, now buried

Gulf of Mexico

N

0   50 km
0   50 miles

YUCATÁN PENINSULA

Cenotes (sinkholes) form ring, due to water dissolving weakened rocks near crater wall

Normal cenote pattern elsewhere

Skull was one of the largest of any land animal and accounted for one third of the animal's 6-tonne body weight

Horns may have been used for defence or in courtship

Smaller third horn, hence its name, triceratops

Rows, or batteries, of ridged teeth could shear through the tough-leaved plants like cycads that flourished at the time

Beaklike toothless jaw tip was probably used to grab vegetation

death of the dinosaurs

In the aftermath of its creation, Earth was bombarded with asteroids. This coincided with the period of life's single-celled origins, but life's birthplace on the ocean floor possibly helped shield it from harm. Since then, most asteroids striking the planet have been too small to have a lasting influence on the biosphere, so life has been free to evolve into extraordinary complexity. But around 66 million years ago, Earth was hit by an asteroid that changed the course of evolution. It scattered dust and soot into the atmosphere and plunged Earth into a dark winter that lasted for thousands of years, driving three-quarters of plant and animal species to extinction – including the giant dinosaurs.

Insects

More than 20,000 insect fossils have been recovered from the Messel Pit, representing all major groups of insects living today. Delicate features such as wings, antennae, and eyes have been preserved. In some specimens the light-reflecting nanostructures of their exoskeletons are still evident, resulting not only in iridescent fossils, but also allowing scientists to reconstruct the vibrant colours of these ancient insects.

Iridescent colours clearly visible

GROUND BEETLE
Ceropria messelense

Delicate hind wing structure has been preserved

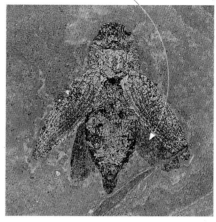

JEWEL BEETLE
Psiloptera weigelti

Cold-blooded vertebrates

The warm, greenhouse climate of the Eocene supported a rich diversity of fish, amphibians, and reptiles. The Messel Lake and surrounding subtropical forests were home to more than 30 species of reptiles, including turtles, lizards, snakes, and crocodiles. Some of the skeletons had been "frozen" in their final moments; there are fossilized pairs of turtles mating and even snakes with their last meals in their stomachs.

Bones of the spiny dorsal fin similar to those of a living perch

FISH
Amphiperca multiformis

Ankle bones are not fused unlike today's frogs

FROG
Eopelobates wagneri

Warm-blooded vertebrates

An incredible range of mammals and birds have been found in the Messel Pit. Early representatives of major animal groups living today show that ecosystems of the Eocene were beginning to resemble modern communities. Hoofed proto-horses mingled with early tapirs, primitive rodents, and tree-living primates. Bats shared the skies with a true aviary: more than 70 species of birds are known from the area.

Long legs typical of wading birds

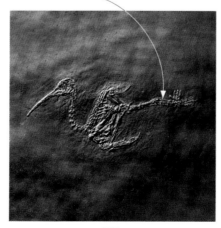

IBIS
Rhynchaeites messelense

Wing membrane seen between upper limb bones

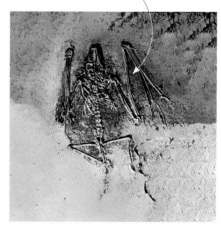

BAT
Palaeochiropteryx tupaiodon

fossilized community

Lagerstätten, from the German word meaning "storage places", are sedimentary deposits around the world in which fossils are exceptionally preserved. One of these, the remains of an ancient volcanic lake in Germany called the Messel Pit, provides an unparalleled window into an Eocene ecosystem from 47 million years ago; more than 53,000 plant and animal fossils have been found. The lack of oxygen and currents at the bottom of the lake ensured preservation, but scientists debate why so many species were fossilized. One explanation is that toxic gases were periodically released from the lake, killing animals in the surrounding subtropical forest. Alternatively, algal blooms may have poisoned animals that drank the water.

Large ovipositor (organ used to deposit eggs) is prominent

PARASITIC WASP
Xanthopimpla messelensis

Domed carapace compacted by sediment

TURTLE
Allaeochelys crassesculptata

Keeled scales help animal move through vegetation

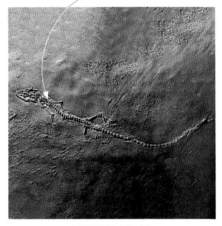

MONITOR LIZARD
Paranecrosaurus feisti

Remains of powerful trunk muscles that constrict prey

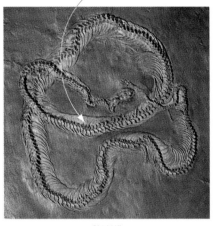

SNAKE
Eoconstrictor fischeri

Jaw musculature similar to that of a guinea pig or porcupine

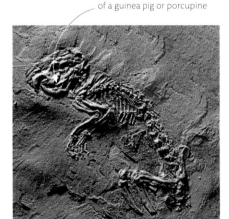

RODENT
Masillamys beegeri

Opposable toes well-suited to climbing

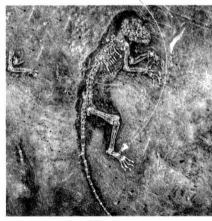

PRIMATE
Darwinius massillae

Rhinoceros-like teeth supports ancestral links to both tapirs and rhinos

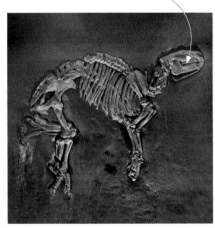

TAPIRLIKE HOOFED ANIMAL
Hyrachyus minimus

the age of mammals

As organisms evolve, they adapt to play out different roles, or niches, within the biosphere. Over 100 million years ago – when dinosaurs still dominated the land – mammals occupied the roles of small nocturnal hunters. After dinosaurs were wiped out (see pp.294–95), live-bearing mammals rose up to fill their niches – diversifying, in the absence of competition, into new kinds of herbivores and carnivores. They diverged into two main groups – marsupials, which nurtured their young in a pouch, and placentals, which nourished young for longer inside the womb. But some evolutionary trajectories converged – including predatory marsupials in Australia that came to resemble placental big cats in other parts of the world.

Vertebrae extended into a long tail – missing in this specimen

Elbow joint allowed considerable rotation of the clawed forelimb, perhaps to help grapple with prey seized by the jaws

Marsupial lion skeleton
Living 2 million years ago, *Thylacoleo carniflex* was one of Australia's top predators. It had the cutting teeth and sharp claws of the big cats, but – as a marsupial – was more closely related to modern kangaroos and wombats. It occupied an ecological niche similar to that of a true cat and may have climbed trees – perhaps to hide its kills, in parallel with the African-Asian leopards of today.

Hind limb was proportionately shorter than that of a true lion, which limited its running speed

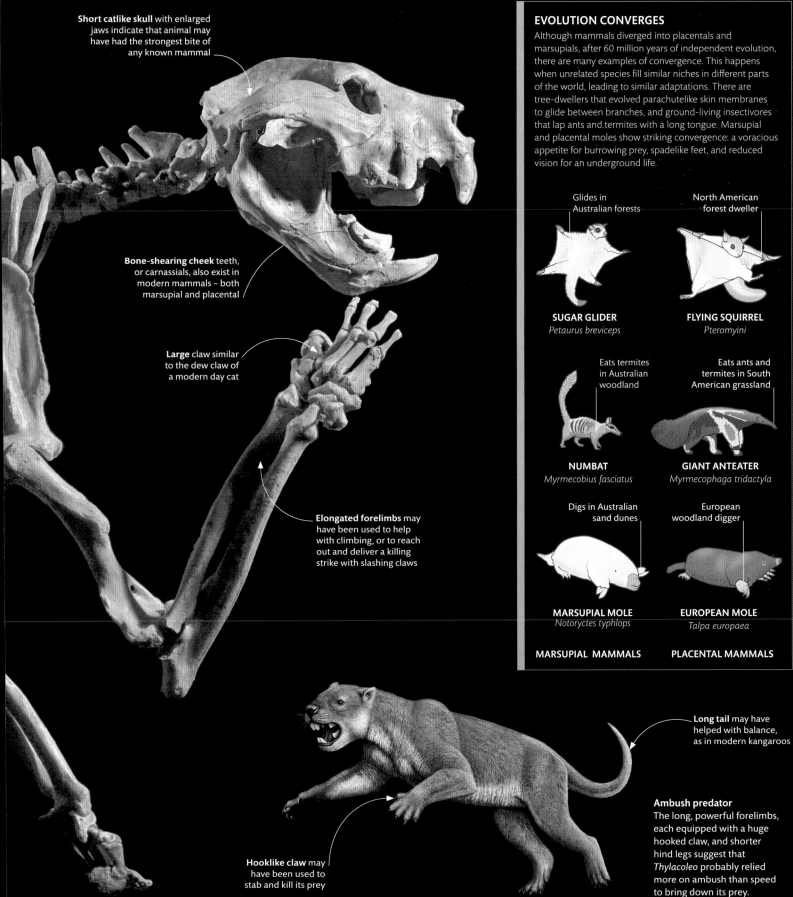

Short catlike skull with enlarged jaws indicate that animal may have had the strongest bite of any known mammal

Bone-shearing cheek teeth, or carnassials, also exist in modern mammals – both marsupial and placental

Large claw similar to the dew claw of a modern day cat

Elongated forelimbs may have been used to help with climbing, or to reach out and deliver a killing strike with slashing claws

Hooklike claw may have been used to stab and kill its prey

EVOLUTION CONVERGES

Although mammals diverged into placentals and marsupials, after 60 million years of independent evolution, there are many examples of convergence. This happens when unrelated species fill similar niches in different parts of the world, leading to similar adaptations. There are tree-dwellers that evolved parachutelike skin membranes to glide between branches, and ground-living insectivores that lap ants and termites with a long tongue. Marsupial and placental moles show striking convergence: a voracious appetite for burrowing prey, spadelike feet, and reduced vision for an underground life.

Glides in Australian forests

North American forest dweller

SUGAR GLIDER
Petaurus breviceps

FLYING SQUIRREL
Pteromyini

Eats termites in Australian woodland

Eats ants and termites in South American grassland

NUMBAT
Myrmecobius fasciatus

GIANT ANTEATER
Myrmecophaga tridactyla

Digs in Australian sand dunes

European woodland digger

MARSUPIAL MOLE
Notoryctes typhlops

EUROPEAN MOLE
Talpa europaea

MARSUPIAL MAMMALS

PLACENTAL MAMMALS

Long tail may have helped with balance, as in modern kangaroos

Ambush predator
The long, powerful forelimbs, each equipped with a huge hooked claw, and shorter hind legs suggest that *Thylacoleo* probably relied more on ambush than speed to bring down its prey.

Horse pulls grass into its mouth with its grasping lips

Grazers on the move

Blue wildebeest (*Connochaetes taurinus*) are dependent upon the African savanna for food, as grass makes up 90 per cent of their diet. Here, in the Masai-Mara, Kenya, they undertake annual migrations in their thousands following seasonal rains that generate the best grazing.

Temperate-zone grazer

On the steppes of Central Asia, the wild Przewalski's horses (*Equus przewalskii*) search for the best grazing in summer, building fat reserves that help sustain them through winter when grass growth stops.

FROM BROWSERS TO GRAZERS

Mammals that graze on low grass descended from browsers (those that eat leaves, shoots, and fruit of taller, woody vegetation) like the *Hyracotherium* and *Mesohippus*. More specialist grazers – such as *Merychippus* and modern horses, asses, and zebras – have higher-crowned, more deeply rooted cheek teeth with self-sharpening ridges. They clip the vegetation with their front incisors, before the cheek teeth grind grass blades as the jaw moves from side to side.

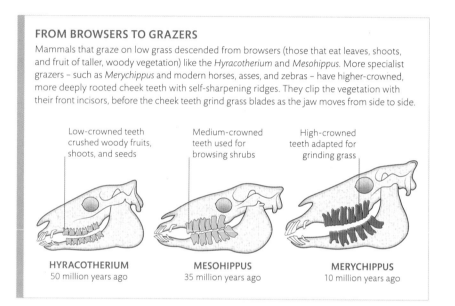

Low-crowned teeth crushed woody fruits, shoots, and seeds

Medium-crowned teeth used for browsing shrubs

High-crowned teeth adapted for grinding grass

HYRACOTHERIUM
50 million years ago

MESOHIPPUS
35 million years ago

MERYCHIPPUS
10 million years ago

grasslands

No single group of plants has had as much bearing upon the nature of the modern biosphere's habitats as grasses. Grasses defend themselves with shards of silica, which makes them tough to eat, and their leaves grow from the base of their blade – so they can quickly regenerate after being "attacked" by any grazing herbivore. These adaptations have also helped to drive their success – especially in the wake of the asteroid collision that sent the dinosaurs and so much other life to extinction (see pp.294–95). In the "new" post-asteroid biosphere when forests waned, the grasses grew so well that they created an entirely different kind of habitat: the open, mainly treeless, steppes and savannahs of a new world dominated by large mammals (see pp.312–13).

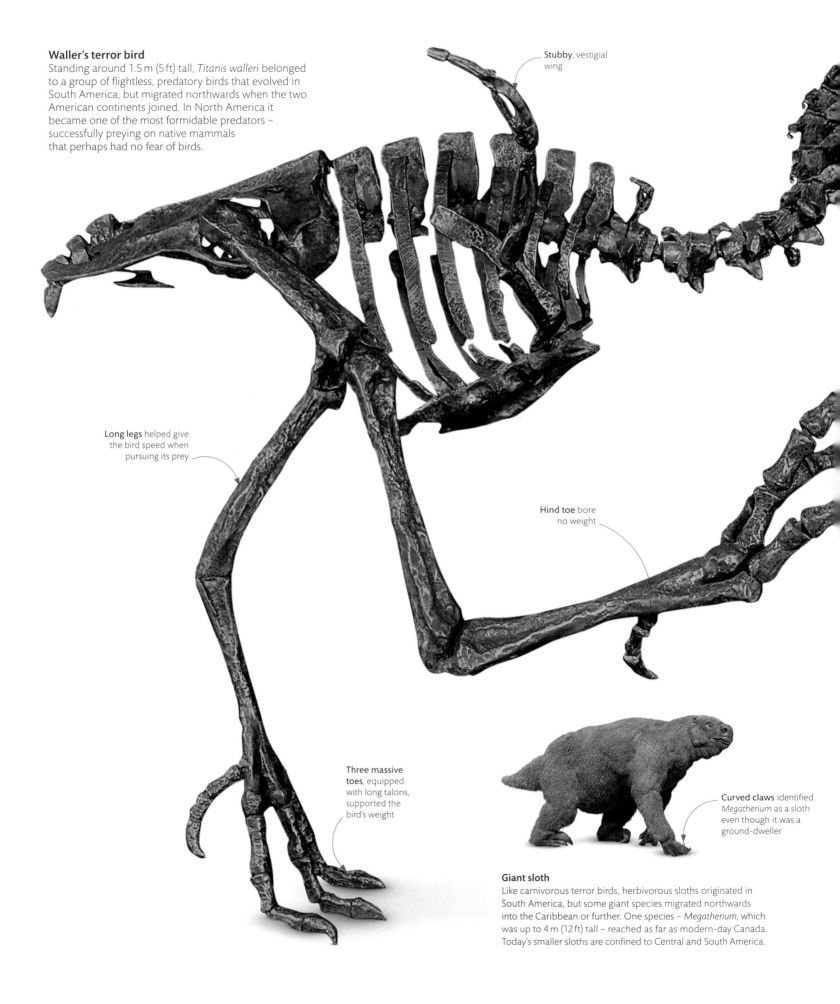

Waller's terror bird
Standing around 1.5 m (5 ft) tall, *Titanis walleri* belonged to a group of flightless, predatory birds that evolved in South America, but migrated northwards when the two American continents joined. In North America it became one of the most formidable predators – successfully preying on native mammals that perhaps had no fear of birds.

Stubby, vestigial wing

Long legs helped give the bird speed when pursuing its prey

Hind toe bore no weight

Three massive toes, equipped with long talons, supported the bird's weight

Curved claws identified *Megatherium* as a sloth even though it was a ground-dweller

Giant sloth
Like carnivorous terror birds, herbivorous sloths originated in South America, but some giant species migrated northwards into the Caribbean or further. One species – *Megatherium*, which was up to 4 m (12 ft) tall – reached as far as modern-day Canada. Today's smaller sloths are confined to Central and South America.

Skull bones were reinforced to support the weight of the large bill

Deep-bladed bill used for gripping and cutting prey

life on the move

Earth's changing geography plays a pivotal role in the evolution of its biosphere. Over millions of years, as continents drifted and islands rose, a dynamic Earth helped to shape animal distributions – such as kangaroos in Australia, sloths in South America, and flightless birds on oceanic islands. When continents split apart and collide, their faunas follow suit. A dramatic meeting of different worlds happened several million years ago, when North and South America came together, and animals living on different continents came face-to-face with counterparts that had been separate since the time of the dinosaurs.

THE GREAT AMERICAN INTERCHANGE

Groups of mammals evolved in isolation on North or South America when the continents were separate. Camels, for instance, evolved in North America, while sloths originated in South America. Both were part of a mammal interchange, first via Caribbean islands, and then across the isthmus that joined the landmasses. Sloths migrated north, and today's guanacos – descendants of early camels – live as immigrants in South America.

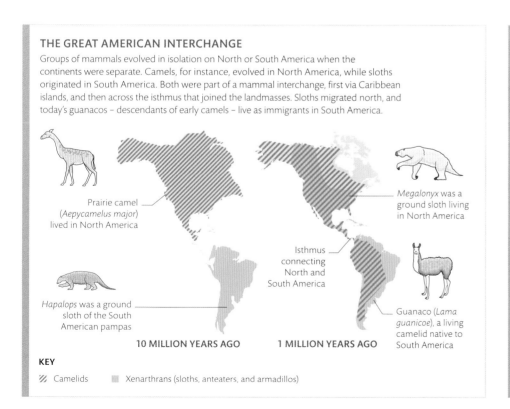

Prairie camel (*Aepycamelus major*) lived in North America

Hapalops was a ground sloth of the South American pampas

10 MILLION YEARS AGO

Isthmus connecting North and South America

1 MILLION YEARS AGO

Megalonyx was a ground sloth living in North America

Guanaco (*Lama guanicoe*), a living camelid native to South America

KEY
▨ Camelids   ▨ Xenarthrans (sloths, anteaters, and armadillos)

ISOLATION ON ISLANDS

Islands form as new volcanic lands erupt from the seafloor or as fragments split away from drifting continents, but new life can evolve in isolation on both. Volcanic islands must be colonized over seas, but without predators or strong competitors, their pioneers can evolve in radical new ways. The large, flightless dodos on Mauritius descended from ancestral pigeons that flew from Asia, but like many island species, evolution left them vulnerable to outsiders. Dodos failed to survive the onslaught of human hunters and went extinct around 1662.

Thick bill used for crushing seeds

Short wing

FLIGHTLESS DODO
Raphus cucullatus

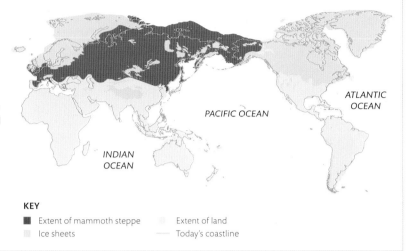

ATLANTIC
OCEAN

PACIFIC OCEAN

INDIAN
OCEAN

MAMMOTH STEPPE DURING THE PLEISTOCENE ICE AGE

KEY

■ Extent of mammoth steppe   Extent of land
  Ice sheets   ------ Today's coastline

surviving ice

Cold is an enemy of life: it slows metabolism, while freezing kills tissues. The coldest places today are at the poles, where the Sun's rays are too feeble to provide much warmth, so life must adapt. But throughout prehistory, tilts in planetary orbit or shifting continents have resulted in parts of Earth being covered with ice. The most recent ice age began just over 2.5 million years ago in the Pleistocene, and its coldest glacial periods (see p.243) saw much of North America and Eurasia under an Arctic ice sheet. It turned the land beyond into freezing steppe, where cold-resistant large mammals grazed where few other species could.

Ice-age rhinoceros
In an example of convergence (see p.299), the woolly rhinoceros (*Coelodonta antiquitatis*) evolved a large body size and shaggy coat of long hair that helped to trap body heat – just like the mammoth's.

Curved tusks found on both males and females were longer and more curved than those of today's elephants

Shoulder hump housed fat, muscle, and well-developed neck ligament that supported head and horn

Ice-age mammoth

Descended from tropical elephants, the woolly mammoth (*Mammuthus primigenius*) is one of the best-known animals to thrive during the Pleistocene Ice Age. Remains of mammoths found preserved in sub-zero temperatures show that, in life, they had a cold-adapted metabolism and were insulated by a subcutaneous fat layer and a thick coat. They also relied more on grazing grass than does any living elephant.

Preserved specimens show that tusks spiralled downwards and outwards

Long, high-domed skull raised the crown of the head, like today's Asian elephant – its closest living relative

Jaws carried molar (cheek) teeth with complex ridges used to grind tough grass and sedges of the steppe

Fossilized skeletons reveal that *Mammuthus primigenius* was the same size as a modern African savannah elephant

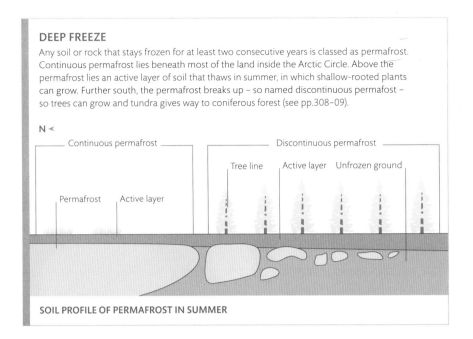

Any soil or rock that stays frozen for at least two consecutive years is classed as permafrost. Continuous permafrost lies beneath most of the land inside the Arctic Circle. Above the permafrost lies an active layer of soil that thaws in summer, in which shallow-rooted plants can grow. Further south, the permafrost breaks up – so named discontinuous permafost – so trees can grow and tundra gives way to coniferous forest (see pp.308–09).

N

Continuous permafrost

Discontinuous permafrost

Tree line   Active layer   Unfrozen ground

Permafrost   Active layer

SOIL PROFILE OF PERMAFROST IN SUMMER

permafrost

At times, Earth's climate has been so warm that even the poles were unfrozen (see pp.242–43). But today the Arctic and Antarctic are both capped by polar ice sheets that wax and wane with the seasons. Around the Arctic circle, however, even in summer, the ground remains frozen beneath a certain depth. The ice blocks the downward growth of roots and restricts the upward growth of vegetation. This is the world of the Arctic tundra – a place dominated by low-growing plants, such as sedges and mosses – where few woody trees and shrubs can survive.

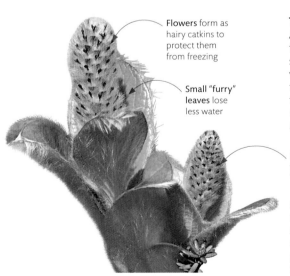

Flowers form as hairy catkins to protect them from freezing

Small "furry" leaves lose less water

Plant rarely grows to more than 15 cm (6 in)

Tundra polygons
Although very little rain falls in the tundra, the low temperatures slow its evaporation, so the land stays wet and permafrost stops water draining away. Water-filled cracks in the ground freeze with wedges of ice, which then thaw during the summer, and the meltwater collects in huge polygonal pools, as seen in this northern Alaskan landscape.

Northern shrub
No woody plant lives further north than the Arctic willow (*Salix arctica*). Despite being a relative of the willow trees of temperate latitudes, this plant is tiny, and it hugs the ground. Its buds are insulated from the bitter icy winds by woolly hairs.

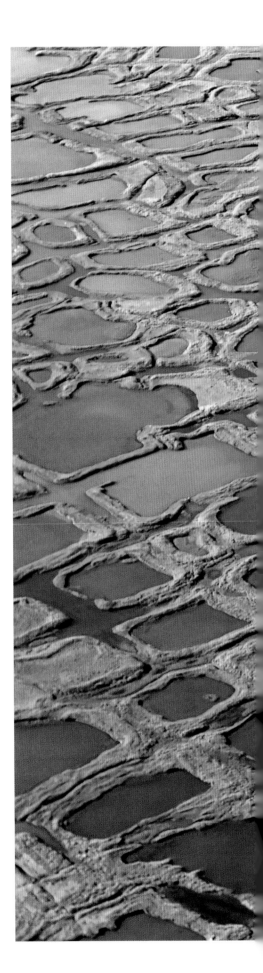

boreal forest

Also known as taiga forest, boreal forest is the biosphere's newest and most extensive natural habitat. Around 12,000 years ago, as the last glaciation (see pp.304–05) was receding, the polar icecaps shrank, uncovering more land in North America and Eurasia. Ice that had long laid buried in soil – the permafrost (see pp.306–07) – melted, and the dry steppes gave way to damper forest. Boreal forest encircles the globe between tundra to its north and broadleaved forest and grasslands further south. But the climate here, close to the Arctic Circle where the Sun stays low above the horizon, is still very cold and the coniferous trees that grow here must be able to survive these conditions.

Forests of the north
Cold-adapted forests, such as this one in Finland, are dominated by needle-leaved, cone-bearing conifer trees. These conifers include evergreen spruces and pines and deciduous larches, species that are hardy enough to survive temperatures that remain below freezing for most of the year.

Circumpolar animals

The boreal forests that stretch across North America and Eurasia provide similar kinds of habitats, so many species of animals, such as the moose (*Alces alces*), range across the entire region.

Antlers occur only in males, and can span 1.2–1.5 m (4–5 ft) across

World's largest living deer browse on willow and birch in moist clearings

---

CONIFER ADAPTATIONS

Foliage of conifers is quite different from that of broadleaved, flowering plants. Their leaves are shaped like needles or reduced to small scales, which reduces their surface area, limiting water loss by evaporation, so the trees can survive when so much ground water is frozen. These adaptations help conifers to grow in tropical, arid parts of the world too. Conifer tree sap also contains resins that work as antifreeze, and the broader leaves have a waterproof waxy coating.

Wide surface area

BROAD LEAF

Short, tightly packed scalelike leaves

Awl-like leaves radiate from a single stem

Pointed-tipped, linear leaves

Needle-like leaves stem from a single bract

TYPES OF CONIFER LEAF

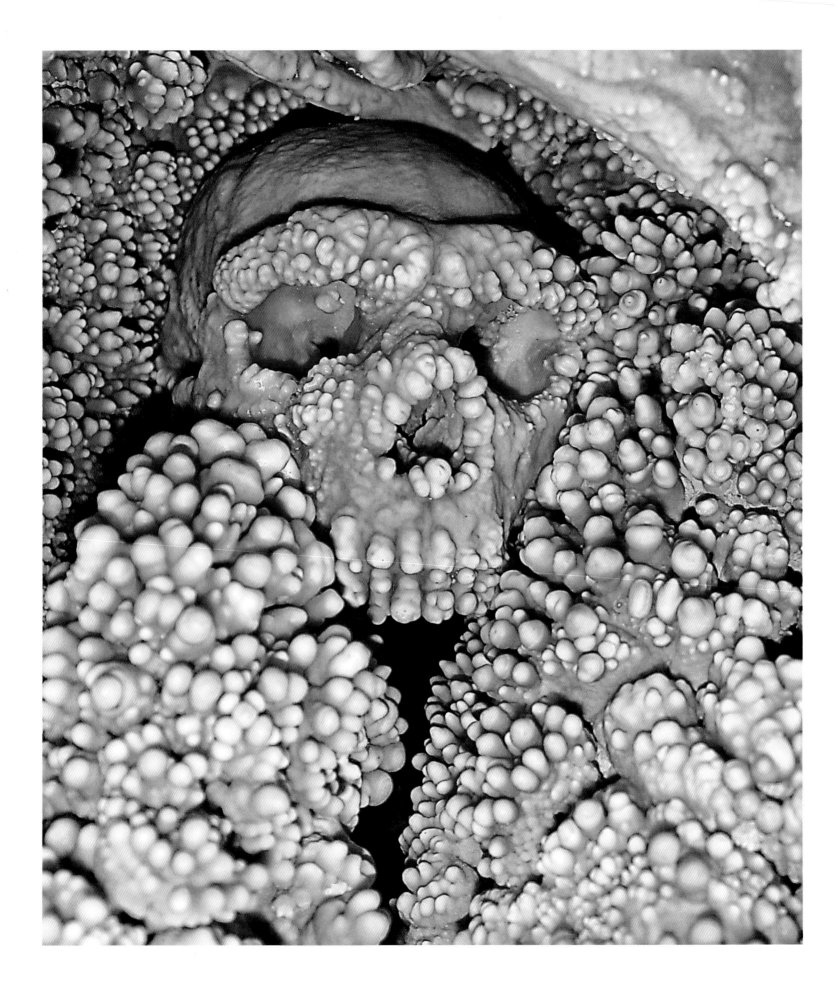

Altamura Man
The fossilized remains of Altamura Man (left), discovered in Italy in 1993, have provided a rich source of Neanderthal DNA. Analysis of DNA from Neanderthal remains shows that they interbred with modern humans – an insight that has revolutionized palaeoanthropology.

" The challenge for the molecular palaeontologist is to discover chemical traces of ancient life. "

DEREK E. G. BRIGGS AND ROGER E. SUMMONS 2014

history of Earth science

molecular evidence from fossils

Fossilized feather remains

SINOSAUROPTERYX FOSSIL

REDDISH MELANOSOMES

BLACK-GREY MELANOSOMES

Exceptionally preserved fossils — including those of "soft" parts that are not usually mineralized — have long provide insight into the shapes and features of extinct organisms. Modern analytical methods are now showing that these spectacular fossils of prehistoric life may preserve more than meets the eye: cellular details, molecular signals, and possibly even fragments of DNA.

The potential preservation in fossils of biological molecules (biomolecules), including proteins, carbohydrates, and genetic material, was first recognized in the 1950s. In the 1970s and 1980s, analysis of biomolecules progressed, and with advances in DNA sequencing, the first DNA from an extinct animal was extracted from the 19th-century remains of the extinct zebra-like quagga. By refining these approaches, scientists working with remains that have been preserved, but not fossilized, have since been able to identify the genetic make-up of many extinct animals, such as human relatives, ice-age horses, and even million-year-old woolly mammoths.

In the mid-2000s, the recovery of original soft tissues from *Tyrannosaurus rex* spurred debate about whether DNA might be preserved in fossils. Remnants of soft tissue have now been found in a wide range of dinosaur fossils. However, scientists still debate the nature of these preserved tissues, and whether it is possible to distinguish between original tissues and those contaminated by the presence of microorganisms or other genetic material.

Further studies have shown that many different kinds of biomolecules, especially proteins, can be transformed during fossilization into stable forms that provide information about the organisms when they were alive. Transformed biomolecules have now been used to study, for example, the relationships of some of the earliest life forms, the colours of dinosaurs, and even to identify fossilized blood in the stomach of an Eocene mosquito (56–34 million years ago). Ongoing advances in these techniques continue to push the boundaries of the study and understanding of ancient biomolecules and life forms.

Ancient melanosomes
Under the electron microscope (left), the melanosomes (microscopic sacs of pigment within cells) of fossilized feathers can be seen. Red melanosomes tend to be spherical, while black ones are sausage-shaped. The fossilized remains of *Sinosauropteryx* (above left) indicate that the creature had feathers with dark and pale stripes.

megafauna

In the last 60 million years, mammals became the new big beasts on Earth – from huge ground sloths in America (see p.303) to giant marsupials in Australia. Like the giant reptiles before them (see pp.290–91), they followed a trend that may have been driven by the benefits of outsizing competitors and predators. In a world that had been cooling since the peak of the dinosaurs, large warm-blooded mammals were good at maintaining body temperature. Today's megafauna is greatly reduced – most giant mammals became extinct within the last 50,000 years, a timing that coincides with the dispersal of humans, suggesting that these formidable hunters were at least partly to blame.

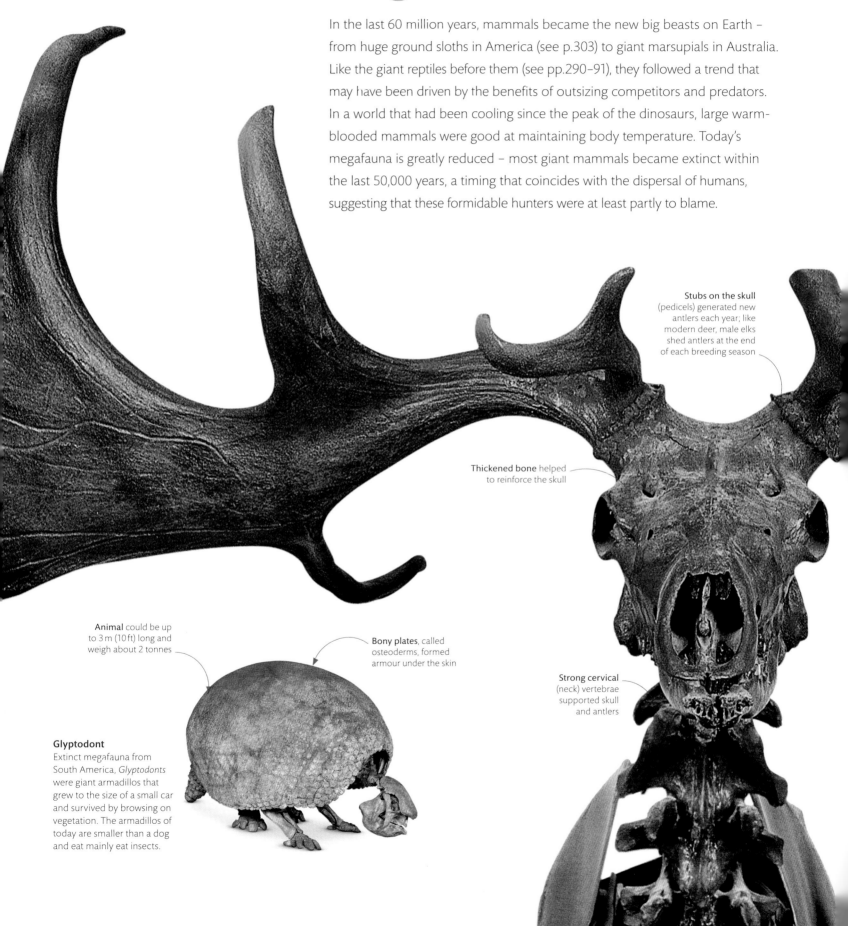

Stubs on the skull (pedicels) generated new antlers each year; like modern deer, male elks shed antlers at the end of each breeding season

Thickened bone helped to reinforce the skull

Animal could be up to 3 m (10 ft) long and weigh about 2 tonnes

Bony plates, called osteoderms, formed armour under the skin

Strong cervical (neck) vertebrae supported skull and antlers

Glyptodont
Extinct megafauna from South America, *Glyptodonts* were giant armadillos that grew to the size of a small car and survived by browsing on vegetation. The armadillos of today are smaller than a dog and eat mainly eat insects.

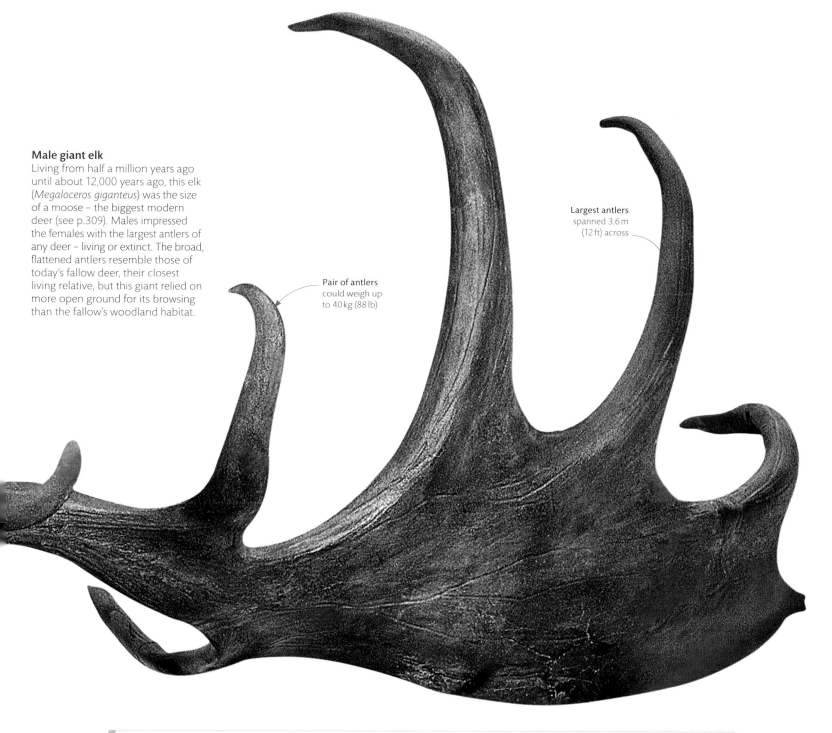

Male giant elk

Living from half a million years ago until about 12,000 years ago, this elk (*Megaloceros giganteus*) was the size of a moose – the biggest modern deer (see p.309). Males impressed the females with the largest antlers of any deer – living or extinct. The broad, flattened antlers resemble those of today's fallow deer, their closest living relative, but this giant relied on more open ground for its browsing than the fallow's woodland habitat.

Pair of antlers
could weigh up
to 40 kg (88 lb)

Largest antlers
spanned 3.6 m
(12 ft) across

MEGAFAUNA DIMENSIONS

The heaviest megafauna, all weighing several tonnes, include now extinct mammoths, as well as living giants such as giraffes and bison, all of which are herbivores: processing tough plant food requires big guts. But the carnivores that hunted them also became huge. Megafauna species numbers reached their peak in the Pleistocene epoch – about 2.5 million years ago – a time that corresponded with the last ice age (see pp.304–05).

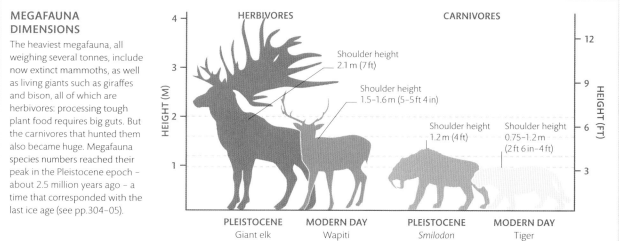

HERBIVORES

CARNIVORES

Shoulder height
2.1 m (7 ft)

Shoulder height
1.5–1.6 m (5–5 ft 4 in)

Shoulder height
1.2 m (4 ft)

Shoulder height
0.75–1.2 m
(2 ft 6 in–4 ft)

HEIGHT (M)

HEIGHT (FT)

PLEISTOCENE	MODERN DAY	PLEISTOCENE	MODERN DAY
Giant elk	Wapiti	*Smilodon*	Tiger

Rising eerily from the beach along Cardigan Bay on the west coast of Wales, the submerged forest of Borth is a reminder of a bygone age. Stretching for 7 km (4 miles) between the towns of Borth and Ynyslas, these preserved trunks of ash, oak, birch, and pine are entombed in a layer of peat some 6,000 years old, which protected them as they were submerged by rising seas. Today, they are sporadically exhumed by storms, which scrape

Forest of Borth

away overlying sand and leave the twisted roots exposed at low tide. Sightings of the prehistoric forest are becoming more commonplace, and more and more of the forest is exposed as warming oceans produce more frequent and powerful storms.

Locals link the trees to the legend of Cantre'r Gwaelod, or the Lowland Hundred – a sunken ancient kingdom. As the story goes, the civilization was inundated after a prince – either forgetful, lustful, or inebriated – left open a sluice gate, allowing the sea to wash over the low-lying area. Scholars are less eager to jump to conclusions. The ephemeral exposure of the trees does make them challenging to study, but the reefs imagined as city walls are more likely to be glacial moraines. Nonetheless, the area was clearly once inhabited as stone hearths, wooden walkways, and stone tools have been found amongst the stumps.

Similar submerged forests are found around the coasts of Great Britain, which was connected to mainland Europe during the last glacial period (around 115,000–11,700 years ago, see pp.304–05). Only when the glaciers receded did sea level rise, flooding the once-fertile land bridge and isolating the island.

Algae-covered tree remains are exposed as overlying sand is washed away

Tree stump and roots at low tide

Landscape of the forest
In 2014 a major storm scoured the shoreline of Ceredigion County and, as the waves receded, the stumps of an ancient forest once buried in peat were exposed. Research shows that the trees stopped growing 4,000–6,000 years ago. The remains are visible at very low tides; they could begin to rot if exposed for long periods, but seawater also has a preservative effect.

shaped by humans

One species, *Homo sapiens*, has had a far greater impact on the rest of the biosphere than any other single kind of living thing. Its entire fossil record is less than half a million years old, but its impact on the environment will last far longer. So great are the effects of humankind that scientists have proposed a new geological epoch – the Anthropocene – to mark its significance. By 10,000 years ago, humans had colonized much of the world, altering landscapes with their agriculture and hunting large animals to extinction. Since then, a new age of industry and technology is polluting the world and changing the climate, causing global warming.

Humanity's mark
Less than one-quarter of Earth's land surface remains as wilderness – largely untouched by humanity. These giraffes in Nairobi National Park, Kenya, stand in a 100 square km (39 square mile) area of protected land – but one that is fenced in on three sides and has Kenya's capital city, Nairobi, looming in the distance.

Plastic crisis
This plastic cup may be a temporary home for a hermit crab (*Coenobita* sp.), but plastics resist decomposition and are ultimately part of a pollution crisis in the ground and oceans that is threatening the natural world.

Discarded cup is a plastic substitute for a natural shell

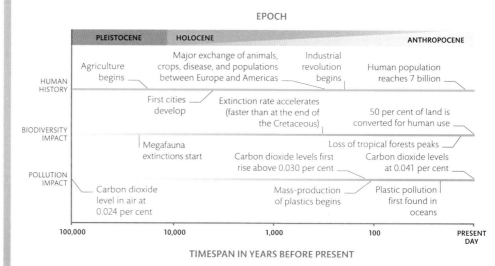

THE GRADUAL ONSET OF THE ANTHROPOCENE

Like all species, humans use resources and generate waste, but they also modify the environment extensively through agriculture, transport, urbanization, and fuel combustion. Forests and other habitats make way for fields and cities, invasive species are introduced, and technology and industry generate pollutants. No single event defines the Anthropocene – experts differ on when it began.

EPOCH

PLEISTOCENE	HOLOCENE		ANTHROPOCENE

HUMAN HISTORY
- Agriculture begins
- Major exchange of animals, crops, disease, and populations between Europe and Americas
- Industrial revolution begins
- Human population reaches 7 billion

BIODIVERSITY IMPACT
- First cities develop
- Extinction rate accelerates (faster than at the end of the Cretaceous)
- 50 per cent of land is converted for human use
- Megafauna extinctions start
- Loss of tropical forests peaks

POLLUTION IMPACT
- Carbon dioxide levels first rise above 0.030 per cent
- Carbon dioxide levels at 0.041 per cent
- Carbon dioxide level in air at 0.024 per cent
- Mass-production of plastics begins
- Plastic pollution first found in oceans

100,000	10,000	1,000	100	PRESENT DAY

TIMESPAN IN YEARS BEFORE PRESENT

spotlight # Danum Valley

Although the Anthropocene (see p.317) has been bad news for nature, pockets of pristine, diverse wildlife habitat do survive. While the tropical rainforest on the island of Borneo is under threat (more than 50 per cent of old-growth forest has been lost since 1973), nestled on the northeastern coast of the island is the Danum Valley Conservation Area – more than 400 square km (150 square miles) of uninhabited rainforest. One of the last expanses of undisturbed old-growth forest left in Southeast Asia, it is dominated by dipterocarp trees, some of the tallest flowering trees in the world. There are 270 dipterocarp species on Borneo, of which 155 exist only on the island, and most, if not all, are found in the Danum Valley. The area is home to unique plant species, like the insect-eating pitcher plants, the world's largest flower – the corpse flower *Rafflesia* – and the tallest tropical tree,

yellow meranti (*Shorea faguetiana*). Inhabiting the forest are populations of critically endangered animals such as Bornean orangutans (*Pongo pygmaeus*), Bornean pygmy elephants (*Elephas maximus borneensis*), and the clouded leopard (*Neofelis nebulosa*). The valley is a sanctuary for more than 124 mammal and 340 bird species vulnerable to hunting or habitat fragmentation.

Biodiversity research in designated conservation areas such as Danum Valley – protected from hunting and logging – demonstrates that conservation efforts work and can successfully guide management policies. As a result, the rate of deforestation has recently decreased, and protected areas have expanded.

Red leaf monkey (*Presbytis rubicunda*)

Adult monkey
weighs only 6 kg (13 lb) when fully grown

Heart of Danum Valley
Morning mist hangs over the rainforest of the Danum Valley in Sabah, Borneo. The dipterocarp trees seen here produce distinctive winged seeds once in a decade, and do so en masse, flooding the jungle with billions of seeds. Many endemic plant and animal species survive here away from the threat of deforestation and hunting.

glossary

Terms that appear in **bold text** refer to other entries in the glossary

ABYSSAL A term applied to the depths of the ocean. The abyssal plain is the almost flat plain of the deep ocean floor beyond the continental margin. The abyssal zone is the zone of water 4,000–6,000 m (13,000–20,000 ft) deep.

ACCRETION The process by which material is added to a continental plate at a **subduction** zone.

AFTERSHOCK A smaller **earthquake** or tremor following a larger earthquake.

ALGA (pl. ALGAE) Any organism that **photosynthesizes** but is not a true plant, including seaweeds and many microscopic forms. **Cyanobacteria** are usually now excluded from the definition.

ALLUVIAL A term applied to sedimentary material deposited by rivers. An alluvial fan is a deposit laid down where a stream leaves a highland area and spreads out over a plain. See also **Sediment**.

AMMONITE The fossilized coiled shell of an extinct cephalopod **mollusc** (a relative of squid) that died out about 66 million years ago. Members of the broader group of animals that includes ammonites are called ammonoids.

AMNIOTE A **vertebrate** whose embryo is protected by membranes and sometimes a shell. Reptiles, birds, and **mammals** are all amniotes.

ANGIOSPERM A flowering plant (as distinct from a **conifer**, fern, moss, or other plant). Angiosperms include many tree species.

ANION A negatively charged ion. See also **Cation**, **Ion**.

ANTICLINE An archlike, upwards **fold** of originally flat strata formed by horizontal compression. See also **Syncline**.

ANTICYCLONE A weather system in which winds circle around an area of high pressure. See also **Cyclone**, **Hurricane**, **Tropical cyclone**.

ARACHNID A member of the group of arthropods that includes mites, spiders, scorpions, and relatives.

ARCHOSAUROMORPH A member of the **clade** of **diapsid** reptiles that includes crocodiles, **dinosaurs**, and birds.

ASTEROID One of thousands of bodies of rocky material orbiting the Sun. Their diameters range from a few metres to about 1,000 km (600 miles).

ASTHENOSPHERE The layer of the **mantle** immediately below the rigid **lithosphere**. It is sufficiently non-rigid to flow slowly in a solid state and plays a key part in the movement of **tectonic plates**.

ATOLL A ring-shaped group of coral islands surrounding a **lagoon**, usually on top of a **seamount**.

AURORA A display of lights in the night sky in Arctic and Antarctic regions, caused by the interaction between electrically charged particles of the Sun and Earth's magnetic field.

AUSTRALOPITH One of several species of upright apes, including *Australopithecus*, living 1.2–4.4 million years ago.

BASALT Earth's most common volcanic rock, which usually originates as solidified **lava**. Basalt is glassy to fine-grained (composed of very small **crystals**).

BATHOLITH A huge, irregularly shaped mass of **igneous rock**, 100 km (60 miles) or more across, formed from the intrusion of **magma** from deep underground. See also **Pluton**.

BEDDING The manner in which a **sedimentary rock** is laid down in layers. A bedding plane is a surface in the rock that separates these layers.

BIOGENIC Produced by the activity of living creatures or biological processes.

BIOMINERALIZATION The process by which living creatures produce minerals.

BIOTIC Relating to, or produced by life and living organisms.

BIVALVE An aquatic **mollusc**, such as an oyster or clam, with two hinged shells that cover its whole body.

BLACK SMOKER A type of **hydrothermal vent** in which the hot water is coloured dark by sulfide minerals. A white smoker emits other minerals, such as silica and barite.

BRACHIOPOD A major group of marine **invertebrates**. Their two-part shell makes them look superficially similar to **bivalve molluscs**, but they are unrelated. They were abundant and diverse in the **Paleozoic** and Mesozoic.

BRACKISH A term to describe water that is slightly salty.

BRECCIA A **sedimentary rock** made up of angular fragments, called **clasts**, cemented by **minerals**. See also **conglomerate**.

BRINE A highly concentrated solution of salt in water, often found as sea water.

CALCAREOUS Containing or made of the **mineral** calcium carbonate.

CALDERA A bowl-shaped volcanic depression larger than a **crater**, typically greater than 1 km (0.6 miles) in diameter, caused by the collapse of a volcano into its emptied **magma** chamber.

CALVING The process by which a **glacier** creates icebergs by shedding ice blocks into a sea or lake.

CAPILLARY WAVE A wave at the boundary of a fluid that is dominated by the **surface tension** of the fluid, rather than gravity.

CARAT A term used to describe the proportion of gold in a gold alloy, with 24-carat gold being pure gold. It is also a unit of weight, equivalent to 0.2 g (0.007 oz), used for diamonds and precious stones.

CARTILAGINOUS FISH A class of fish whose skeleton is made of cartilage, not bone. It includes sharks and rays.

CATION A positively charged ion. See also **Anion**, **Ion**.

CHEMOSYNTHESIS The process by which an organism can grow and multiply using the energy stored in simple chemicals, such as hydrogen sulfide or methane. Chemosynthesis functions in contrast to **photosynthesis**, which depends on energy coming from the Sun. Many bacteria can carry out chemosynthesis, notably those living around **hydrothermal vents**.

CINDER CONE A relatively low-lying volcano formed from cinders and ash that have fallen from an eruption cloud.

CLADE A group of species consisting of all the evolutionary descendants of a given ancestor.

CLAST A fragment of **mineral** or rock, especially when incorporated into newer, **sedimentary rock**.

CLEAVAGE The way some **minerals** break along preferred planes (known as cleavage planes) dictated by their atomic structure.

CLIMATE CHANGE Long-term changes to weather patterns and average temperatures on a global or regional scale.

COMET A body composed of rock and ice that orbits the Sun, usually with a highly eccentric orbit. As it approaches the Sun, vaporization of the ice produces an extensive cloud, called the coma, and one or more tails.

COMPOUND A substance containing atoms of two or more **elements**.

CONDENSATION NUCLEUS A tiny particle in the air around which a raindrop or snow crystal may form. See also **Precipitation**.

CONGLOMERATE A **sedimentary rock** that is made up of rounded, gravel-size **clasts** cemented together by **minerals**. See also **Breccia**.

CONIFER A cone-bearing tree. Nearly all conifers, such as pines and firs, have needle-like leaves and are evergreen. Some, such as larch, lose their needles in the winter months.

CONTACT METAMORPHISM The processes by which rocks are transformed by the effects of contact with hot **magma**. See also **Dynamic metamorphism**, **Regional metamorphism**.

CONTINENTAL CRUST The part of Earth's **crust** that forms the continents. It is less dense and thicker than **oceanic crust**.

CONTINENTAL ISLAND A body of land surrounded by water that lies on the **continental shelf** of a continent.

CONTINENTAL RISE The region between deep ocean seabed and the **continental slope**.

CONTINENTAL SHELF The relatively flat and shallow seafloor surrounding a continent, which is considered geologically part of that continent. See also **Continental slope.**

CONTINENTAL SLOPE The sloping seabed that leads from the edge of a **continental shelf** down to the **continental rise**.

CONTOUR CURRENT A slow-moving current that flows parallel to the **continental rise**.

CONVECTION Circulating currents (for example, in Earth's atmosphere or **mantle**) caused by differences in temperature. This results in the warmer, less dense air or fluid rising upwards, which drives the circulation.

CONVERGENT BOUNDARY The line along which two **tectonic plates** that are moving towards each other meet.

CORE Earth's innermost layer. It consists of a liquid outer core and a solid inner core, both made of nickel-iron. See also **Mantle**, **Crust**.

CORIOLIS EFFECT The tendency for winds and currents moving in a northerly or southerly direction to be deflected and move at an angle because of the effect of Earth's rotation. The deflection is to the right in the northern hemisphere and to the left in the southern hemisphere.

CRATER A bowl-shaped depression through which an erupting volcano discharges gases, rock fragments, and **lava**. The crater walls form by accumulation of ejected material. Also refers to a circular depression in the landscape caused by a large meteorite impact. See also **Caldera**.

CRATON A stable area of Earth's **continental crust** that is made of old rocks largely unaffected by mountain-building activity since the Precambrian. Also called a shield.

CROSS-BEDDING Layering in **sedimentary rock** that is at an angle to the main bedding plane. See also **Bedding**.

CRUST Earth's outermost rocky layer. The continents and their margins are made of thicker but less dense **continental crust**, while thinner, denser **oceanic crust** underlies the deep ocean floors. See also **Mantle**, **Tectonic plate**.

CRYSTAL Any solid in which the individual atoms or molecules are arranged in a regular geometrical pattern compared to unordered solids, such as glass. There are seven basic patterns of crystal growth, called crystal systems. See also **Geode**, **Habit**.

CYANOBACTERIA An abundant group of microscopic **photosynthesizing** bacteria, formerly called blue-green **algae**. Their tiny cells lack nuclei, as do those of other bacteria.

CYCAD A member of the group of non-angiosperm seed plants that are superficially similar to palms, but produce seeds and pollen in cones. See also **Angiosperm**.

CYCLONE A pressure system in which air circulates around an area of low pressure. See also **Anticyclone**, **Hurricane**, **Tropical cyclone**.

CYNODONT A member of a group of advanced **synapsids** that arose in the Late Permian. **Mammals** are their living descendants.

DELTA The area of gently sloping **sediment** built up by a river when it enters the sea, a lake, or another river. Its shape depends on the sediment carried by the river, and the currents and tides it meets. See also **Estuary**, **Submarine fan**.

DENDRITIC A term applied to **minerals** occurring naturally in a branching, tree-like form within rocks.

DEPOSITION The laying down of material such as sand and gravel in new locations, usually by wind, water, or ice.

DESERT PAVEMENT A rocky or stony surface layer found in many deserts.

DIAPSID A major division of **amniotes** whose name refers to the two openings on each side of the skull. Traditionally regarded as reptiles, diapsids include the **archosauromorphs** (including **dinosaurs**, birds, and crocodilians) and lepidosaurs (including lizards and snakes). See also **Synapsid**.

DICYNODONTS A group of herbivorous **synapsids** (the ancestors of **mammals**) with two tusks and a blunt beak.

DIFFERENTIATION A process in which heavy **elements** in a body of **magma**, such as the early Earth or a magma chamber, sink while lighter elements float.

DIKE A sheetlike intrusion of **igneous rock** that cuts across existing rock structures.

DINOSAUR A major group of **archosauromorph amniotes** that had an erect gait with walking limbs directly beneath the body. They became extinct at the end of the Cretaceous, with the exception of their descendants, birds.

DIVERGENT BOUNDARY The line along which two **tectonic plates** are moving away from each other.

DIVERSIFICATION Multiplication of species over evolutionary time, often in the process of filling ecological niches.

DOWNWELLING The sinking down of water from the surface of the ocean. Large-scale downwelling in certain regions gives rise to a process known as **thermohaline circulation**. See also **Upwelling**.

DYNAMIC METAMORPHISM A type of metamorphism in which a rock is transformed by pressure being applied in a particular direction during large-scale movement of Earth's crust. See also **Contact metamorphism**, **Regional metamorphism**.

EARTHQUAKE A sudden shaking of Earth's surface caused by **seismic waves** travelling through the **lithosphere**.

ECHINODERM A member of the marine **invertebrate** phylum that includes starfish, sea urchins, sea cucumbers, and relatives. These animals feature tube feet used for feeding and movement.

ECOSYSTEM The entirety of living and non-living features of a region and the interactions between them.

ECTOTHERMIC A term to describe a "cold-blooded" animal, one that regulates its body temperature through external heat sources, such as sunlight.

ELEMENT A substance that cannot be broken down into simpler substances.

ENDOTHERMIC A term to describe a "warm-blooded" animal, one that regulates its body temperature through internal metabolic processes.

EPICENTRE The point on Earth's surface directly above the point of origin of an **earthquake**. See also **Seismic**.

EROSION The processes by which rocks or soil are loosened and worn or scraped away from a land surface. The main agents of erosion are wind, moving water, and ice, and the rock particles that they carry.

ESTUARY A broad, tidal, funnel-shaped stretch of **brackish** water found where many large rivers meet the sea.

EXTRUSIVE ROCK A rock formed from **lava** that flowed onto Earth's surface or was ejected as **pyroclastic** material.

FAST ICE A continuous sheet of **sea-ice** that is frozen onto the coast, or land-formed ice at its landward edge.

FAULT A fracture where the rocks on either side have moved relative to one another. If the fault is at an angle to the vertical, and the hanging wall block (above the fault plane) has slid downwards, it is called a normal fault. If the hanging wall block has slid upwards (in relative terms) it is called a reverse fault. A strike-slip fault is one where movement is horizontal.

FELDSPAR A common type of **silicate** mineral found in **igneous rock**.

FETCH The extent of open water across which a wind or water wave has travelled.

FIORD A former glacial valley on the coast that has become an inlet of the sea.

FLOODPLAIN A flat plain next to a river that is liable to be covered with water when the river floods.

FOLD A geological structure in which originally flat-lying layers of rock have been bent under compression. They may bend upwards in the middle to form a ridge (**anticline**) or downwards to form a trough (**syncline**). In an overturned fold, one side (or limb) of the fold moves more than the other and extends over the other limb.

FOLIATION The arrangement of **minerals** in parallel bands in some deformed **metamorphic rocks**.

FOSSILIFEROUS A term applied to **sedimentary rocks** containing fossils.

FROST SHATTERING **Weathering** caused by the alternate freezing and thawing of water in cracks in a rock.

FULGURITE Naturally formed tubes of glassy **minerals** formed when lightning strikes the ground.

FUMAROLE In volcanic regions, a small opening in the ground through which steam and hot gases can escape.

GASTROPODS The largest class of **molluscs**, which includes all snails and slugs, as well as limpets and sea slugs.

GEMSTONE A jewel or stone used in jewellery that is prized for its durability, beauty, and rarity.

GEODE A hollow cavity within a rock that is lined with **crystals**.

GEYSER A jet of boiling water and steam that rises at intervals from the ground. It is powered by hot rocks heating the surrounding **groundwater**.

GINKGO A seed-bearing tree with fan-shaped leaves that is native to China.

GLACIER A mass of ice that flows slowly downhill from an **icecap** or mountainous region. A valley glacier flows between the walls of a valley. A tidewater glacier is a valley glacier that flows into the sea.

GONDWANA An ancient supercontinent that included South America, Africa, Antarctica, Australia, and India.

GORGE A narrow, deep valley, usually with vertical cliffs on either side.

GRAPTOLITE Member of an extinct group of colonial and mostly planktonic **invertebrates**, typically growing in the form of long narrow colonies of small individuals, supported by a skeleton.

GREASE ICE The first stage of formation of **sea-ice**. Ice **crystals** form in the sea water, giving it a thick, oily texture.

GROUNDWATER Water occurring below the surface of the land and held in spaces (interstices) between the grains of rocks. The upper limit of such a groundwater zone is called the **water table**.

HABIT The external shape of a **crystal** (or group of crystals), including the forms and faces of individual crystals.

HADROSAUR A member of the group of duck-billed herbivorous **dinosaurs** that lived during the Late Cretaceous.

HANGING VALLEY A valley, usually carved by a **glacier**, that enters high up the side of a larger valley.

HARDNESS Of a **mineral**, the extent to which it can resist scratching or abrasion. See also **Mohs scale**.

HEADWATER The upper portion of any river or stream, close to its source.

HORSETAIL A group of spore-bearing plants whose above-ground stems have a jointed appearance, with rings of small branches bearing tiny scalelike leaves. Some extinct forms were as big as trees.

HOTSPOT A long-lived zone of volcanic activity thought to originate deep in Earth's **mantle**. **Tectonic plates** passing over hotspots are marked by linear chains of volcanoes that become progressively older with increased distance from the hotspot. See also **Mantle plume**.

HUMUS A dark-coloured substance found in soils, derived from dead plants, microorganisms, and animals.

HURRICANE A huge, circular tropical storm in which there are wind speeds of 119 kph (74 mph) or more. It is also called a **tropical cyclone** and (especially in east Asia) a **typhoon**. Its energy comes from the latent heat of water that has evaporated from warm oceans and later condenses.

HYDROTHERMAL VEIN A thin, sheetlike body of minerals, laid down by the circulation of hot, mineral-bearing waters in Earth's **crust**.

HYDROTHERMAL VENT A fissure in a volcanically active area of the ocean floor, where water rich in chemicals issues from the rock at high temperatures. See also **Black smoker**, **Chemosynthesis**, **Hydrothermal vein**.

ICECAP A dome-shaped mass of ice that submerges a landscape from a single mountain to an entire polar region.

ICE FLOE A large area of floating **sea-ice**.

ICE SHEET A very large, flowing mass of ice that permanently covers a land area, as in Antarctica or Greenland.

ICE SHELF A floating extension of an **ice sheet** or **glacier** over the ocean.

IGNEOUS INTRUSION A body of **igneous rock** that formed when **magma** cooled and solidified underground. See also **Batholith**.

IGNEOUS ROCK Rock that originates from the solidification of molten **magma**. See also **Metamorphic rock**.

INTRUSIVE ROCK **Igneous rock** that has solidified below the surface, cooling slowly enough to allow larger **crystals** to form. A body of intrusive igneous rock is called an **igneous intrusion**.

INVERTEBRATE An animal without a backbone, such as an insect, snail, or worm. See also **Vertebrate**.

ION An atom or group of atoms that has lost or gained one or more electrons to become electrically charged. The molecules of most minerals include a combination of **cations** (positively charged ions) and **anions** (negatively charged ions).

ISOSTATIC REBOUND The rise of a land mass after the huge weight of an **ice sheet** is removed.

ISOTOPE One of two or more forms of the same chemical **element** that have equal numbers of protons but different numbers of neutrons in the nuclei of their atoms.

KARST A characteristic landscape formed of water-soluble rocks, especially limestone, that have been weathered by water flowing underground. A karst coast occurs where limestone has been weathered and later inundated by the sea.

LAGOON An area of sheltered sea water, almost cut off from the open ocean.

LARVA (pl. LARVAE) The young stage of an animal when it is very different in structure from an adult.

LAURASIA An ancient supercontinent that included North America, Europe, and most of Asia (apart from India).

LAVA Molten rock (**magma**) that flows onto Earth's surface.

LIGHT MICROGRAPH A photograph taken through a microscope that uses visible light.

LITHOSPHERE The layer of Earth that includes the **crust** and the upper **mantle**.

LOBE-FINNED FISH A mainly extinct group of fish having fleshy, muscular bases to their paired front and back fins. They are first known from the Silurian period and include the present-day lungfish and coelacanth.

LOESS An unconsolidated sedimentary deposit made up of wind-blown silt. One of the largest examples is the Loess Plateau in northern China. See also **Sedimentary rock**.

LUSTRE The way a **mineral** reflects light and the extent of its sheen.

MACROCRYSTALLINE A term applied to **crystals** that are large enough to be recognized with the naked eye. See also **Microcrystalline**.

MAGMA Liquid, molten rock in Earth's **mantle** and **crust**. It cools to form **igneous rock**. It may crystallize beneath the surface or be erupted as **lava**.

MAGNETOSPHERE The region around Earth (or other planet) dominated by the planet's magnetic field.

MAMMAL A class of warm-blooded **vertebrates**, almost all of which give birth to live young and nourish them using milk produced by the female.

MANTLE The rocky layer of Earth between the **crust** and the **core**. It makes up 84 per cent of Earth's volume.

MANTLE PLUME A column of hot rocks rising through the **mantle** and **crust**, giving rise to a **hotspot** at Earth's surface.

MARGINAL SEA A partially enclosed sea bordering a continent.

MARSUPIAL A **mammal**, such as a koala or kangaroo, in which offspring are born in a relatively undeveloped state; they often continue their growth within an external pouch on the mother. See also **Placental**.

MASSIF A well-defined mountainous region whose rocks and landforms tend to be similar across the region.

MASSIVE Of a rock: having a structure that does not show layering or other divisions into smaller segments.

MATRIX The mass of relatively fine-grained material that binds together larger particles in some heterogeneous **sedimentary rocks** and larger **crystals** in volcanic rocks. See also **Porphyritic**.

MESOSPHERE The layer of the Earth's atmosphere between the **stratosphere** and **thermosphere**, at an altitude of about 50–80 km (30–50 miles).

METAMORPHIC AUREOLE A zone of rock around a body of **magma** that has been altered by **contact metamorphism**.

METAMORPHIC ROCK A rock that has been transformed underground by heat or pressure, altering its texture or minerals. For example, marble is metamorphosed limestone. See also **Igneous rock**.

METEOR A small mass of rock from space that vaporizes completely as it falls through Earth's atmosphere, glowing as it does so.

METEORITE A rock from space that has fallen to Earth's surface without completely burning up.

METEOROID A potential **meteor** or **meteorite** before it has entered Earth's atmosphere.

MICROCRYSTALLINE Having **crystals** that are so minuscule that they are detectable only with the aid of a microscope. See also **Macrocrystalline**.

MID-OCEAN RIDGE An undersea mountain range running along the deep ocean floor, and a site where new oceanic **crust** is created.

MINERAL Any solid, naturally occurring inorganic material that has a characteristic **crystal** structure and well-defined chemical composition. Most rocks are mixtures of more than one mineral.

MOHS SCALE A measure of the **hardness** (resistance to scratching) of minerals, ranging from 1 to 10.

MOLLUSC A member of the group of soft-bodied **invertebrates** that includes **gastropods**, **bivalves** and cephalopods (such as octopuses and squid). See also **Ammonite**, **Nautiloid**.

MONOCOT Also called a monocotyledon, a member of the group within the **angiosperms** that includes grasses, orchids, palms, and lilies. They are characterized by having only one cotyledon (seed-leaf) within each seed.

MONOTREME An egg-laying **mammal**, such as the present-day platypus and echidna (spiny anteater). Egg-laying is thought to be the original mode of reproduction for **mammals**. See also **Marsupial**, **Placental**.

MORAINE An accumulation of rock debris resulting from glacial action. Active valley **glaciers** create lateral moraines at their edges, medial moraines where two glaciers merge, and terminal moraines at their ends. Moraines often remain after a glacier has melted. See also **Outwash**, **Striation**.

MOSASAUR A member of an extinct group of Cretaceous marine reptiles that are thought to be relatives of snakes and monitor lizards.

MULTISPECTRAL Relating to two or more wavelengths in the electromagnetic spectrum. Multispectral imaging can be used to make maps of natural and geological features.

MULTITUBERCULATE A member of an extinct group of mainly rodentlike early **mammals** found from the Jurassic to the Paleogene period.

NATIVE ELEMENT A chemical **element** that occurs in a pure, or uncombined, state in nature.

NAUTILOID A member of the group of cephalopod **molluscs** related to the **ammonites** and nautiluses. Their coiled external shell contains gas-filled chambers that lighten the body in the water.

NEAP TIDE A tide in which the difference between high and low tide is at its smallest. See also **Spring tide**, **Tidal bore**.

NODULAR About a rock: containing rounded lumps of minerals or other materials. For example, flint in chalk.

NON-CRYSTALLINE A term to describe a solid material that does not have a consistent arrangement of particles.

NUTRIENTS Chemicals, especially salts of **elements** such as nitrogen, phosphorus, and iron, that are essential for living organisms to grow.

OCEANIC CRUST The part of Earth's **crust** that underlies most of the world's oceans. It is thinner and denser than **continental crust**.

ORE A rock from which metal can be profitably mined.

ORNITHISCHIAN A member of one of the two main subgroups of **dinosaurs** (the other being saurischians), the name literally meaning "bird-hipped". Ornithischians include ornithopods, stegosaurs, ankylosaurs, and ceratopsians.

OUTWASH The deposit of sand, gravel, and other materials carried by the meltwater from a **glacier**. See also **moraine**.

OZONE A form of oxygen with three atoms in its molecules. It is present in the upper atmosphere (the ozone layer), where it absorbs ultraviolet radiation.

PALEOZOIC The era of geological time extending from the beginning of the Cambrian to the end of the Permian (539–252 million years ago). It was followed by the Mesozoic.

PANGAEA An ancient supercontinent that included almost all the present-day continents.

PARAREPTILE A member of a diverse group of extinct **amniotes** traditionally called reptiles. They include mesosaurs and several other groups.

PHOTOSYNTHESIS The process by which plants, **algae**, and **cyanobacteria** use the energy of sunlight in the presence of chlorophyll to convert water and carbon dioxide into food. See also **Chemosynthesis**.

PHYTOPLANKTON The tiny, mostly single-celled organisms that live in the surface waters of oceans and lakes, and are the foundation of most aquatic food chains. See also **Algae**, **Plankton**.

PILLOW LAVA Pillow-shaped mounds of rock formed from **lava** that has been ejected underwater.

PLACENTAL Those **mammals** in which the foetus grows to a relatively advanced stage in the mother's womb, nourished by a placenta. Includes all living mammals except **marsupials** and **monotremes**.

PLACODERM A group of extinct jawed fish, widespread in the Devonian period, whose skin was protected by bony armour-plating.

PLANETESIMAL One of millions of rocky objects of variable size believed to have been present in the early history of the Solar System, and which later came together to create the planets.

PLANKTON Any living species (plants, animals, or microorganisms) living in open water that drift with the currents. Most plankton are small, but some – such as jellyfish – are larger. See also **Phytoplankton**, **Zooplankton**.

PLANKTON BLOOM A rapid increase in the population of **phytoplankton** in oceans or lakes. It can cause the water to appear blue-green, brown, or even red.

PLUME The column of ash, volcanic particles, and gas released during an explosive volcanic eruption.

PLUNGE POOL The deeper area at the bottom of a waterfall formed by the force of water hitting soft bedrock at its base.

PLUTON A smaller mass of **igneous rock** that has formed beneath the surface of Earth by solidification of **magma**. See also **Batholith**.

PORPHYRITIC A term to describe an **igneous rock** texture in which large **crystals** are set in a finer **matrix**.

POTHOLE A hole formed in the bed of a river by the action of water carrying small pebbles and **sediment**.

PRECIPITATION Water that reaches Earth's surface from the atmosphere, including rain, snow, hail, and dew.

PRIMATE A member of the group of **mammals** that includes monkeys, apes, and humans. Typical characteristics include grasping hands and forward-facing eyes.

PRISMATIC A term to describe **crystals** in which the parallel rectangular faces form prisms.

PROTOZOA Animal-like single-celled organisms, mainly microscopic, that are common in virtually all habitats and may be free-living or parasitic. The thousands of species are in many different subgroups, not all closely related. See also **Zooplankton**.

PTEROSAUR A flying **diapsid** related to the **dinosaurs**, with wings formed by skin stretched over their front limbs. Originating in the Triassic, they went extinct at the end of the Cretaceous.

PYROCLASTIC Consisting of, or containing, volcanic rock fragments. Pyroclastic flows are fast-moving, sometimes deadly clouds of hot gases and debris.

RAIN SHADOW An area of low rainfall downwind of a mountain range, caused by air shedding moisture as it passes over the windward side of the range.

RAISED BEACH A beach that is now above high-tide level because of a change in sea level or uplift of the land.

RAVINE A smaller valley with steep sides. See also **Hanging valley**, **Rift valley**.

RECRYSTALLIZATION The transformation of smaller **crystals** into larger ones that takes place under very high pressure during metamorphism.

REFRACTION The change of direction of waves, including light waves, when they pass from one medium to another. Water waves change direction when they reach shallow water.

REGIONAL METAMORPHISM The transformation of rocks by metamorphism over broad areas, such as in mountain ranges.

RHYNCHOSAUR A member of the group of extinct **archosauromorph** herbivores. One of the most common reptiles of the Triassic.

RIFT VALLEY A large block of land that has dropped vertically downwards compared with its surroundings as a result of horizontal extension of the crust and normal faulting. See also **Fault**.

ROGUE WAVE An unexpected and unusually large ocean wave.

SALINITY The concentration of dissolved salts in, for example, water or soil.

SALT A **compound** formed from the reaction of an acid with a base. Also, the common name for sodium chloride.

SALT LAKE A body of landlocked water containing water that is very salty because of evaporation.

SALT PAN A shallow depression, often in a desert, that contains salt deposits.

SAUROPOD A member of the group of huge herbivorous **dinosaurs** (including *Diplodocus* and *Brachiosaurus*) with long necks and long tails that were the largest land animals ever to have lived.

SEAFLOOR SPREADING The process by which new oceanic **crust** is created as **tectonic plates** split apart from one another. See also **Mid-ocean ridge**.

SEA-ICE Ice that arises from sea-water freezing. See also **Fast ice**, **Grease ice**, **Ice floe**, **Ice shelf**.

SEAMOUNT An undersea mountain, usually formed by volcanic activity. See also **Atoll**.

SEA SCORPION An extinct aquatic predatory arthropod, also called a eurypterid. Sea scorpions occurred from the Ordovician to the Permian.

SEA STACK A tall column of rock rising out of the sea near a coastline, a resistant remnant of surrounding cliffs that have otherwise disappeared due to **erosion**.

SEDIMENT Particles carried by moving water, wind, or ice, or deposits formed by such particles, including gravel, sand, silt, or mud. See also **Sedimentary rock**.

SEDIMENTARY ROCK Rock formed when small particles are deposited – by wind, water, volcanic processes, or mass movement – and later compacted and cemented together.

SEED FERN A member of one of the several groups of extinct seed plants that have fernlike leaves despite not being related to ferns.

SEISMIC Relating to **earthquakes**. A seismic wave is a shockwave generated by an **earthquake**.

SEISMOGRAPH An instrument for recording **seismic** waves. A seismogram is a recording made by a seismograph. See also **Earthquake**.

SEISMOMETER The part of a **seismograph** that responds to ground motions, often caused by **earthquakes**.

SHIELD VOLCANO A volcano with shallow slopes that is built from thin, fluid **lava**. See also **Cinder cone**, **Stratovolcano**.

SILICATE Any rock or **mineral** composed of groups of silicon and oxygen atoms in chemical combination with atoms of various metals. Silicate rocks make up most of Earth's **crust** and **mantle**.

SILICEOUS About a rock: containing or consisting of **silicate**.

SILL A roughly horizontal, sheet-like **igneous intrusion** that usually forms when **igneous rock** forces its way between layers of existing **sedimentary rocks**.

SINKHOLE A depression in the surface of a **karst** landscape. It often leads down to an underground drainage system.

SOLAR WIND The stream of charged particles released from the upper atmosphere of the Sun.

SPRING TIDE A tide that occurs when the effects of the Moon and Sun reinforce each other, producing the highest high tide and the lowest low tide. See also **Neap tide**.

STALACTITE A deposit of calcium carbonate hanging down from the roof of a cave or underground passage.

STALAGMITE A deposit of calcium carbonate rising up from the floor of a cave or underground passage.

STRATIGRAPHY The geological study of the order and relative position of strata (layers of **sedimentary rock**) in Earth's **crust**.

STRATOSPHERE A layer of Earth's atmosphere extending from the top of the **troposphere**, at 8–16km (5–10 miles), up to about 50km (30 miles). See also **Mesosphere**.

STRATOVOLCANO A conical volcano that is built from alternate layers of **lava** and other rock fragments produced in an explosive eruption. See also **Cinder cone**, **Shield volcano**.

STRIATIONS Grooves and scratches that are left in the bedrock as a **glacier** moves over them.

SUBDUCTION The descent of an oceanic **tectonic plate** under another plate when two plates converge. Subduction zones can be classified as either ocean–ocean or ocean–continent depending upon the nature of the two converging plates. See also **Crust**, **Convergent boundary**.

SUBMARINE FAN **Sediment** at the base of the **continental rise** that has been deposited on the **seafloor**.

SUBSOIL The layer of soil immediately beneath the **topsoil**.

SURF The tops of waves as they break on rocks or the shore. See also **Swell**.

SURFACE TENSION An effect that makes a liquid seem as though it has an elastic "skin." It is caused by cohesive forces that pull molecules at the surface of a body of water inwards or sideways.

SWELL A series of regular waves created by weather systems over a long distance.

SYNAPSID A major group of tetrapod **vertebrates** that branched off early in the evolution of **amniotes**,

and eventually gave rise to the **mammals**. Synapsids were the largest terrestrial creatures during the Permian. See also **Diaspid**.

SYNCLINE A downwards **fold** of strata that were originally flat, resulting from an application of horizontal compression. See also **Anticline**.

TABULAR A term to describe the **habit** of a **crystal** with predominantly large, flat, parallel faces.

TECTONIC PLATE Any of the large rigid sections into which Earth's **lithosphere** is broken. The relative motions of different plates leads to **earthquakes**, volcanic activity, continental drift, and mountain-building. See also **Convergent boundary**, **Divergent boundary**, **Transform boundary**.

TERRESTRIAL Relating to Earth; found on land rather than in the sea.

TERRIGENOUS A term to describe ocean **sediments** that were created by **erosion** on land.

THERMOCLINE Any layer at a particular depth in an ocean where the average temperature changes rapidly with depth. Thermoclines can also occur in the atmosphere.

THERMOHALINE CIRCULATION A worldwide circulation of deep-water currents that is driven by differences in temperature and **salinity** between different water masses. See also **Coriolis effect**.

THERMOSPHERE The layer of Earth's atmosphere above the **mesosphere**. It extends over altitudes of about 80–640km (50–400 miles).

THEROPOD A member of a major group of bipedal **dinosaurs** that includes top predators such as *Tyrannosaurus rex*, as well as many smaller species, such as *Velociraptor* and the ancestors of modern birds.

TIDAL BORE A single large wave sometimes created when a rising tide enters a narrowing channel such as an **estuary**.

TOPOGRAPHIC Relating to the features and forms of a land surface.

TOPSOIL The outermost layer of soil, containing **minerals** and organic matter in which plants grow. See also **Subsoil**.

TRACTION The process by which coarse **sediment** grains are moved along a surface by water or wind currents.

TRAIN A succession of waves of similar wavelength spaced at regular intervals. See also **Refraction**.

TRANSFORM BOUNDARY A boundary between **tectonic plates** where the plates slide horizontally past each other.

TRILOBITE A member of an enormously diverse group of extinct marine arthropods. Trilobites went extinct at the end of the Permian period.

TROPICAL CYCLONE A large-scale circulating weather system in tropical and subtropical regions that is powered by warm ocean water and produces violent winds and heavy rain. Also known as a **hurricane** or **typhoon**. See also **Cyclone**.

TROPOPAUSE The boundary between the **troposphere** and the **stratosphere**. Air temperature starts to increase with height above the tropopause, which varies from about 16km (10 miles) at the Equator to 8km (5 miles) over the poles.

TROPOSPHERE The lowest, densest layer of the atmosphere, where most weather phenomena occur. The height of the upper limit of the troposphere (known as the **tropopause**) varies from the equator to the poles.

TSUNAMI A fast-moving, often destructive sea wave generated by **earthquake** activity. It rises in height rapidly as it reaches shallow water. See also **Tidal bore**.

TURBIDITY CURRENT A dense flow of **sediment**-laden water in lakes, and from the continental margin down onto the ocean floor.

TYPHOON A **hurricane** or **tropical cyclone** occurring in the western Pacific or Indian oceans.

UNCONFORMITY A gap in the geological record indicating that one or more strata have been removed by erosion. See also **Sedimentary rock**.

UNIFORMITARIAN THEORY The theory that the laws and processes that operate on Earth today also operated in a similar way in the geological past.

UPWELLING The rising of deep ocean water to the surface. It can be caused by wind blowing parallel to a coastline or by an underwater obstruction such as a **seamount** interrupting a deep-sea current. Upwelled water often enriches the surface of the ocean with nutrients. See also **Downwelling**.

VERTEBRATE A backboned animal. Vertebrates include fish, amphibians, and **amniotes**, which include reptiles, birds, and **mammals**. See also **Invertebrate**.

VISCOSITY Resistance to flow in fluids. The higher the viscosity of a fluid, the more sluggishly it flows.

VITREOUS A type of **mineral lustre** resembling that of glass.

WATER TABLE The upper surface of the **groundwater** zone in places where it is not confined by impermeable rocks. Water soaking into the ground will tend to flow downwards until it reaches the water table.

WEATHERING The in-situ breaking down of rocks through contact with ice, water, wind, heat, chemicals, or other agents. See also **Erosion**.

ZOOPLANKTON Animals and animal-like organisms living in the plankton. See also **Plankton**, **Phytoplankton**.

index

Page numbers in **bold** refer to main entries.

67P/Churyumov-Gerasimenko 28–9

A

Aberystwyth Grits 93
abrasion, glacial 153, 170
abyssal red clay 205
accretion 14
acid rain 276
actinolite 90
adaptation
   to aridity **281**
   to cold 304, 305, 306, 308
   conifers **309**
   convergent 299
   seasonal 24–5
Advanced Spaceborne Thermal
      Emission and Reflection
      Radiometer (ASTER) 109
aeolian erosion 151
aerofoils 288
African Plate 120
aftershocks 128
Agassiz, Louis 241
agate 48, 79
age, geological **32–3**, 93, 119
aggregates 38, 39
agriculture 317
air pressure 27, 218, 219
air sacs 290
Alaska (US) 132–3
albatrosses 202–3
albite 50
algae
   aquatic 263
   bare rock 265
   biofilms 251
   blooms 81, 179, 182, 297
   marine 182, 193, 208, 246
   oxygen production 253
   single-celled 250
alluvial fans 156–7
Alps 123, 170–1, 174–5
Altamura Man 310–1
altitude 247
altocumulus clouds 222
altocumuluslenticularis clouds 222, 223
altostratus clouds 222
aluminium 47
amazonite 50
amber 59, **60–1**, 269, 292
amethyst 48, 57, 58
ammonites 56, 286, 294
amorphous carbon 44
amphibians 280, 296
amphiboles 89
amplitude (seismic waves) 132

Anak Krakatau (Indonesia) 264–5
Andes 120
andesite 73
angular unconformities **93**
animals
   age of fish **260–1**
   biosphere 247
   Cambrian explosion **256–7**
   circumpolar 309
   desert **280–1**
   dinosaurs **290–1**
   distribution **302–3**
   diversity of species 249
   DNA 311
   evolution 254
   flight **288–9**
   fossils 268–9, 282–3, 286–7, 311
   icy habitats **304–5**
   isolated on islands **303**
   on land **274–5**
   mammals **298–9**
   megafauna **312–3**
   ocean 198, 206, 256
   ocean islands 208
   open ocean waters 202–3
   oxygen-fuelled giants **272–3**
   rainforest 266, 319
   respiration 247
   sources of heat **284**
   succession 265
   warm-blooded **284–5**
   water 98
Annapurna Massif 125
anorthite 51
Antarctic ice sheet 240–1
Antelope Canyon (Arizona, US) 76–7
antennae 292
Antennae galaxies 12
anthracite 61
Anthropocene epoch **317**, 318
anticlines 119
anticyclonic tornadoes 234
antimony 46
antlers 312–3
Apollo missions 20, 21
apophyllite 54–5
aquamarine (mineral) 39, 58
arachnids 278
Archaean eon 32
Archaeopteryx 288–9
arches, sea 149, 195
Arctic Circle 25, 306, 308
Arctic ice sheet 304
Arctic willow 306
arcus clouds 223
arêtes 153
argon 210, 241
arkose 77
armadillos 312
arsenic 44

artesian wells 224
arthropods 272–3, 274, **278–9**
ash
   deposits 33, 74, 75, 179
   volcanic eruptions 140, 141, 237, 276
asperitas clouds 222
Assynt (Scotland) 84–5
Asteroid Belt 16
asteroids 14, 16, 17, 18, 20, 21, 23, 276, 295, 300
asthenosphere 107, 111, 112, 115, 142
Atacama Desert (Chile) 144–5
atmosphere 20, 22, 28, **210–1**
   aurorae **212–3**
   biosphere 246–7
   circulation 27, **215**
   clouds **222–3**
   ice core data 240–1
   imaging **216–7**
   lightning **236–7**
   natural climate change **242–3**
   oxygenation of 244, **252**, 253, 272
   rain **226–9**
   snow **230–1**
   sprites, jets, and elves **238–9**
   thunderstorms **232–3**
   tornadoes **234–5**
   tropical cyclones **220–1**
   water cycle **224–5**
   weather systems **218–9**
   wind **214–5**
atolls 209
atoms 36
aurorae **212–3**
Australian Shield 110–1
avalanches **102–3**
axis, Earth's 21, 24, 25, 243

B

bacteria 32, 81, 246, 247, 251, 253, 263
Baikal, Lake (Russia) 100–1
Baltic Sea 182–3
Baltistan Peak (K6) (Pakistan) 102–3
banded texture 87
banded-iron 81
bark, tree 270
basalt 30, 54, 64–5, 73, 90, 107, 111, 120, 139, 206
basaltic lava 72
basement rocks 95
batholiths 135
bats 288, 296
beaches 195, 196
bees 292
beetles 292–3, 296
bentonite 40
Bentonite Hills (US) 32–3
Berann, Heinrich 201
Bertrand, Marcel-Alexandre 122, 123

beryl 39
biodiversity
   Cambrian oceans **256**
   Great Barrier Reef 258
   human impact on 317
biofilms **251**
biogenic particles 204, 205
biomineralization 60
biomolecules 311
biosphere 225, 244, **246–7**, 249
   evolution of 303
   impact of humans on 317
birds 288–9, 290, 296
   flightless 302, 303
bismuth 40–1, 44
Black Sea 226–7
black smokers 206
block lava 72
Blyde River (South Africa) 149
body fossils 268
body waves 132
bolts, lightning 237
bomb cyclones 218
bones, fossils 268, 283
boreal forests 306, **308–9**
bores, tidal 186
Borth, forest of (Wales) **314–5**
boundaries, tectonic plates 65, 111, **112–7**, 119, 128, 139
brachiopods 256
brackish water 162
Brahmaputra, River 125
branching, plants 263
breakers 188
breccias 16, 65, **79**
   clast-supported 79
   matrix-supported 79
   volcanic 137
breezes 214
broad leaves 309
bromeliads 266
Broomistega 282–3
browsers (animals) 300
Buache, Philippe 85
bubbles
   in ice 240–1
   in lava 21
   in magma 57
burial metamorphism **86–7**
butterflies 248–9
bytownite 51

C

calcarenite 77
calcite 57, 82, 90, 281
calcium 256
calcium carbonate 60, 77, 79, 82, 204, 205
calcium-rich feldspars 50

calderas 139
calving (glaciers) 175, 193
Cambrian explosion 32, **256–7**, 278
Cambrian period 259, 278, 279
camels 303
Canary Islands 142
Cantre'r Gwaelod (Lowland Hundred) 314
canyons 68, 94–5
  oceans 184, 198, 201
Cape Disappointment (Washington State, US) 188–9
capillary waves 188
caprock 161
carats 59
carbon
  long carbon cycle **271**
  natural forms of 42–3, **44**
  radiometric dating 28
  sea water 179
  soil capture 97
carbon dioxide
  emissions 271, 317
  greenhouse gases 210, 276
  in ice cores 241
  photosynthesis 247
  trapped in oceans 179, 180
carbonates 167
Carboniferous period 271, 272, 273, 274, 288
Carlsbad contact twins 36
carnelian 48
carnivores 290, 298, 313
carnivorous plants 318
carpels 292
cartilaginous fish 260
cascades *see* waterfalls
Cathedral Gorge (Nevada, US) 74–5
caves **168–9**
  cliffs 149
  ice 175
  karst 167
  limestone 82
celestite 57
cells, first living 250
cementation 53, 63, **78–9**
cenotes 168, 295
ceratopsians 286
chalcedony 48, 252
chalcopyrite 52, 54
chalk 82, 246
Chang Heng 131
characteristics, species 249
chemical deposition 63, **80–1**
chemical sedimentary rocks 81
chemical weathering 154
chert 81
chevron folds 119
chiastolite 91
Chicxulub Crater (Mexico) 32, 295
chlorophyll 25
chondrites 16
choppiness (sea conditions) 188
chromium 47
chrysocolla 40
chrysoprase 49
cinder cone volcanoes 139

circulation, ocean **180–1**, 185
cirques 153, 170
cirrus clouds 219, 223
cities 317
citrine 48
cladoxylopsids 266
clastic sedimentary rocks 81
clasts 79
clay 74, 79, 157
cliffs
  river 63, 95, 158
  sea 149, 195
climate
  Moon and 21
  ocean circulation 180
climate change
  fossil fuels 271
  glaciers 172, 175
  human activity 317
  ice sheet data 241
  monitoring 217
  natural 217, 241, **242–3**
  Permian eruption 276
  and sea level 196
clints 167
clouds
  rain 228
  tornadoes 234
  types of **222–3**
  water cycle 224
  weather systems 218, 219
coal 60, 81, 85, 286
  formation of **270–1**, 272
coastal erosion **195**
coastline **194–5**
coccolithophores 204, 205
Coelodonta antiquitatis 304
cold-blooded animals 284, 285, 296–7
colliding plates 89, 94, 112, **114–5**, 119, 120, 172, 208
Colorado Plateau (US) 94, 95, 137
Colorado River (US) 94–5
combustion 271
comets 14, 18, 20, 21, 23, 28
common ancestor 249
compaction 79
complex islands 208
compounds, metal 46, 47
condensation 28, 176, 223, 224, 228, 230, 233
condensation nuclei 223, 226
cone-bearing plants 263, 292, 293
cones, volcanic 139
confluences, glaciers 170
conglomerates **78–9**
conifers 293, 306, 308, **309**
Conocoryphe 278
conservation 317, 318, 319
contact metamorphism **90–1**, 130
contact twins 36
continent-continent collisions 114, 115
continental crust 30, 65, **67**, 107, 111, 112
  colliding plates 114, 115
  igneous rocks **66–7**
continental islands 208
continental plates 111, 116, 120

continental shelves 66, 196, 198
continental slopes 184–5
continental transform boundaries 112
continents
  colliding 114, 115, 303
  formation of **30–1**, 32
  separation of 303
contour currents 198
convection currents 111
convergent boundaries 62, 112, **114–5**
Cooksonia 263
Copericus, Nicolaus 27
copper 38, **44–5**, 46, 47, 52
coral 59, 60, 61, 258, 259
coral atolls 209
coral reefs 198, 201, 208, 209, 249, 256
  formation of **258–9**
coral sands 195
core 22, 30
  inner 107
  outer 107, 213
  of planets 22
  of stars 12
Coriolis effect 27, 180, 182
Coronosphaera mediterranea 204
Cothan marble 250–1
countershading 202
cracks
  cave formation 168
  ice 101
  karst landscapes 167
craters
  impact **16**, 18–9, 20, 21
  Mars 22–3
  ocean floor 184
  volcanic 139, 141, 228
cratons 30
creeks 195
Cretaceous period 285, 286, 288, 292, 295
Cretoparacucujus 292–3
crevices, glaciers 193
crinoids 256
crops 317
crust, Earth's **16**, 22, 28, 65, 70, 87, 107, 111, 137
  continent formation 30
  metals **47**
  minerals 34
  mountain-building **120–1**
crustaceans 274, 278
crystalline cavities **56–7**
crystals
  habits **38–9**
  magma 70
  refraction and internal reflection **41**
  snowflakes 231
  structure **36–7**
  systems 36
  twinning **36**
CT (computerized tomography) scanning 283
cumulonimbus clouds 220, 222, 233, 234
cumulus clouds 222, 223

cumulus congestus clouds 226–7
cuprite 38
currents
  lake 165
  ocean 176, 180, 184, 185, 198, 205
  river 157
  upwelling 182
cyanobacteria 60, 182, 251
cycads 292, 293
cyclones 27, **218–9**
  tropical **220–1**
cyclosilicates 50
Cymatophlebia 288
cynodonts 282, 283
Cyprus 208

D

dacite 73
Danum Valley (Borneo) **318–9**
daylight 25
days **25**
Deccan Traps (India) 295
decomposition 246, 247, 270, 272
deer 313
deflation 150
deforestation 317, 319
deltas 157, **162–3**, 205
  inland 162
  submarine 195
Dent de Morcles (Switzerland) 118–9
deposition 33, 63, 80–1, 104, 156–7, 170
desert pavements **150**
deserts 74, 150–1, **280–1**
  salt cycle 179
Devil's Tower (Wyoming, US) 134–5
Devonian period 260, 266, 278
dew point 223
diagenesis **53**
diamond 40, 42–3, 44, 58
diatoms 182, 205
Dickinsonia 254–5
Dicranurus 278
differentiation 28, 30
dikes
  sheeted basalt 65
  volcanic 135, 137
dinoflagellates 182
Dinosaur Provincial Park (Canada) 286
dinosaurs 32, 285, **290–1**
  birds descended from 289, 290
  end of 276, **294–5**, 298, 300
  fossils 283, 286–7, 311
  lifestyle from footprints **291**
diorite 112
Diplodocus 291
dipterocarps 318, 319
divergent boundaries 65, 112, **116–7**, 165
DNA, fossil 274, 311
dodos 303
dolerite 90, 91
dolomite 57, 81, 167, 168
Dolomites (Italy) 120–1
Doppler shift 217
downpours 226–7

downwelling 182
dromaeosaurs 286
drought 97, **280–1**
dry caves 168
dumortierite 38
Dunkleosteus 260–1
dust particles 223, 230
dust storms 151
dust walls 215
dynamic metamorphism **87**, 90

E

earthquakes **128–9**, **132–3**
  avalanches 102
  colliding plates 115
  faults 126
  measuring **130–1**, 132
  plate boundaries 112
  seismic waves 107, 130–2
  subduction 66
  uplifted beaches 196
  water and 225
  waterfalls 161
earthworms 274
East Pacific Rise 117
eccentricity (Earth's orbit) 243
eclogite 106–7
Ediacaran period 254
eggs
  flowers 292
  hard-shelled 280, 281
El Capitan (Yosemite, US) **68–9**
El Niño 217, 243
El Tatio (Atacama, Chile) 144–5
electrical charge, lightning 237
electromagnetic spectrum 109
electromagnetism 131
electrons 41, 213
elements
  metals **46–7**
  native **44–5**
  origin of 12
elephants 304, 305
elk, giant 312–3
Elrathia 278
elves (TLE) 238
emerald 59
Encrinurus 279
endothermic animals 284
Enhanced Fujita scale 234
environment, human impact on 317
Eocene epoch 296, 297, 311
Eomaia 274–5
epicentre 128, 132
epilimnion 165
epiphytes 266
equinoxes 24
erosion
  coastal **195**
  fossils 268, 269
  glacial **170**
  ice **152–3**, 157, 170
  igneous rocks 66, 67, 137
  karst landscapes 167
  layers of rock 92, 93

erosion *continued*
  mountain building 120–3
  rock cycle 62, 63, 104
  sedimentary rocks 74
  water 95, **148–9**, 157, 158
  waterfall formation **161**
  waves 149, 188
  wind **150–1**, 157
  see also weathering
eruptions, volcanic 62, 63, 70, 107,
    139, **140–1**, 161, 237, 241
  Permian eruption **276**
  types of **141**
Escher von der Linth, Arnold 123
eskers 175
estuaries **162**, 186
Eurasian Plate 114, 117, 120, 124
evaporation 62, 101, 179, 224, 226,
    281, 306, 309
evaporites 81
Everest, Mount 124, 125
evolution 244
  Cambrian explosion **256**
  common ancestor 249
  convergence **299**
  death of the dinosaurs 294–5
  first animals 254
  first land animals 274
  first land plants **262–3**
  first trees 266
  fish **260–1**
  geographical separation **249**
  grazing animals
  Great Dying 276
  mammals 285, **298–9**
  through geographical separation
    249
  and water supply 280
exoskeletons 274, 278, 288
exosphere 210
exothermic animals 284–5
extinction
  death of the dinosaurs 276, **294–5**,
    300
  Great Dying 32, **276–7**
  hunting 303, 317
extrusive igneous rocks 62, 70
eye of the storm 220
eyes, compound 292, 293
Eyjafjallajökull (Iceland) 136–7

F

faces, crystals 38
Fairweather Range (US) 172
fast ice 190
fat, body 284, 305
fault blocks 126
fault lines 116, 128
fault movement 126
fault planes 87, 126
faults 87, **126–7**, 132, 165
feathers 288, 289, 311
feldspars 36, 47, 48, **50–1**, 77
ferns 266
fertilization 292

fetch 188
fiords 172, 195
fish **260–1**, 296
fissure eruptions 141
fissures 54, 70, 97, 116, 117, 206
FitzRoy, Mount (Patagonia)
  66–7
flame clams 198
flies 292
flight **288–9**
floodplains 74
floods 149, 158, 175, 195
flowers
  first **292**, 293
  permafrost zones 306
fluorite 36, 59
focus (earthquakes) 128, 132
fog 222
folds **89**, **118–9**, 120
foliation **89**
food chains 246, 247, 263
footprints 268, 269, 274, 291
foraminifera 205
forest fires 272
forests
  boreal **308–9**
  coal formation 270
  coniferous 306
  diversity of species 249
  formation of **266–7**
  loss of 317
  submerged 314–5
  see also rainforests
forms, crystals 38
fossil fuels 270–1
fossils 244
  biodiversity 249
  crystals in 56
  dinosaur **286–7**, 290
  early mammals 285
  foraminiferan and radiolarian
    microfossils 205
  fossilization **268–9**
  and geological age 93
  and geological mapping 85
  insects 269, 292
  Messel Pit **296–7**
  molecular evidence from **310–1**
  oldest 32
  organic minerals 60, 61
  reefs 259
  scanning **282–3**
  seafloor 255, 256
  trilobites 278–9
Foucault, Jean Bernard Léon 26–7
freeze-thaw weathering **155**
fresh water 162
frogs 296
frost flowers 101
frost-shattering **155**
Fuego-Acatenango Massif (Guatemala)
  140–1
fulgurites 237
fumaroles 145
Fundy, Bay of (Canada) 186
fungi 263, 270
fur 285, 305

G

gabbro 50, 65, 107
Galápagos Islands 209
galaxies 12–3
Ganges, River (India) 125, 195
garnet 58, 106, 107
gemstones **58–9**
geo-stationary orbits 217
geodes 54, **56–7**
geological mapping **84–5**
geological time 28, **32–3**
geophysical satellites 109
geothermal features **144–5**
  hot springs **146–7**
Germanophyton 263
geyserites 146
geysers 54, 145, **146**
gills 260
giraffes 316–7
glaciation 67, 69, 95, **152–3**, 154,
    170–1, 243, 308
Glacier Bay (Alaska, US) **172–3**
glaciers 153, 157, 158, 170–1, 172,
    193, 205, 225
  erosion and deposition 62, **170**
  ice cores 241
  meltwater **175**
  and sea levels 196
glass, natural 237
gliders 288, 289, 299
global warming 175, 190, 196, 210,
    271, 276, 317
Glyptodonts 312
gneiss 30, 89
Gobi Desert (Mongolia) 280–1
gold 44, 47, 179
  white 54
gorges 120, 161
Gorgosaurus 286
Gorner Glacier (Switzerland) 170–1
Grand Canyon (US) **94–5**
granite 48, 66–7, 68, 91, 161
granite intrusions 39, 54
graphite 42, 44
grasslands **300–1**, 308
graupels 237
gravel bars 157
gravitational force fields 186
gravity
  Moon **21**, 24, 186
  Solar System 14
  Sun 23
  tectonic plates 111
gravity potato 109
grazers 300–1
grease ice 190
Great American Interchange **303**
Great Bahama Bank 184–5
Great Barrier Reef (Australia) 258–9
Great Dying 32, **276–7**, 278
Great Plains (Canada/US) 234
Great Rift Valley 116, 165
greenhouse Earth 243
greenhouse gases 210, 217, 241,
    243, 276
Greenland ice sheet 241

Grentz Glacier (Switzerland) 170–1
greywacke 77
Grinnell Glacier (Montana, US) 90–1
Gros Morne National Park (Canada) 107
groundwater
  acidic 168
  frozen 309
  geothermal features 145, 146
  heated 54
  karst landscapes 167
  lakes 165
  oxygen-rich 52
  rivers 158
  soil 97
  water cycle 224
grykes 167
guanacos 303
Guettard, Jean-Étienne 85
Gulf of Alaska 172
Gulf Stream 180
gullies 74, 75, 92
seafloor 184, 185
gypsum 81
gyres 180

H

Haapalops 303
habitable zone 23
habitats
  Anthropocene epoch 318
  boreal forests **308–9**
  desert **280–1**
  forests **266–7**
  grasslands **300–1**
  ice **304–5**
  islands in ocean 208
  loss of 317
  open ocean waters 202–3
  permafrost **306–7**
  rainforest 318–19
  reefs **258–9**
  succession 264–5
habits, crystal **38–9**
haboobs 215
Hadean eon 32
Hadley cells 215
hadrosaurs 286
haematite 253
hailstones 233, 237
Half Dome (Yosemite, US) **68–9**
halite 38, 81
halo, polar 213
Hawaiian Islands 30, 72, 142, 209, 228
headlands 195
headwater 158
hedgehogs 285
Heezen, Bruce 201
Heim, Albert 123
heliocentrism 27
helium 12, 14
hematite 40
herbivores 291, 293, 298, 300, 313
  ice age 304–5
hermit crabs 317

Himalayas 114, 120, **124–5**
hinge zones 119
Hokkaido (Japan) 128–9
Holocene epoch 317
holoplankton 182
Homo sapiens **316–7**
hornfels 90, 91
horns, dinosaurs 295
horns (glacial landform) 153
horses 300
Horsethief Canyon (Canada) 286–7
hot springs 54, 145, **146–7**, 206
hotspots **142–3**, 209
Hudson Bay (Canada) 193
human activities 102, 210, 271, 303, 317
humans, early 310–1
humus 97
hunting 303, 317
Huntoniatonia 278
hurricanes 214, **220–1**
hydrogen 12, 13, 14, 179
hydrothermal veins **54**
hydrothermal vents **206**, 246, 250
hypolimnion 165
hypothalamus 284
Hyracotherium 300

I

ibiscs 296
ice 98, **100–1**
  erosion **152–3**, 170
  meltwater 175
  sea **190–1**
  sediment transport/deposition 157
ice ages 168, 170, 172, 241, 243, 304–5, 308, 313, 314
icecaps 139, 196, 243, 308
ice core analysis **240–1**
ice crystals 222, 223, 230, 237
ice floes 190
ice fronts 172
ice sheets 153, 193, 225
ice shelves **193**
icebergs **192–3**
icecaps 139, 196, 243, 308
icehouse Earth 243
Iceland 63, 65, 117, 138–9, 142, 146–7, 158–9, 161, 190–1, 212–3, 236–7
icicles 161
igneous rocks **64–73**
  continents **66–7**
  extrusive 62, 70
  intrusive 90–1, **134–5**, 137
  lava **70–3**
  oceans **64–5**
  rock cycle 62–3
  textures **70**
incised meanders 158
Indian Plate 114, 124
industry 317
infrared radiation 109, 210, 217
inorganic minerals 246, 247
inorganic sedimentary rocks 81
inorganic solids 36

inosilicates 50
insectivores 299
insects 274, 278, 288
  fossils 269, 292
  giant **272–3**
  Messel Pit 296–7
  pollinators 248–9, **292–3**
interglacials 243
interior, Earth's **107**
interstellar space **12–3**, 14
intrusions 54
intrusive igneous rocks 62, 90–1, **134–5**, 137
invasive species 317
inversion layer, trade wind 228
invertebrates 246, 265
ions 36
iridium 44
iron 12, 17, 22, 44, 47, 107
iron oxide 79, 252, 253
ironstone 286
islands
  isolated species **303**
  in the ocean **208–9**
  reef-fringed 201
  volcanic 30, 208, **209**, 303
  volcanic chains **142**
isostasy 120
isostatic rebound 196

J

jade 59
jasper 49, 57, 79
jawed fish 260–1
jawless fish 260
jet 61
jet stream 216–7, 218
jets (TLE) 238
Jupiter 23
Jurassic period 33, 288

K

kamenitzas 167
Kamokuna lava delta (Hawaii) 30–1
kangaroos 303
Karakoram Mountains 125
Kármán Line 210
karst landscapes 155, **166–7**, 168
Kati Thanda-Lake Eyre (Australia) 80–1
Kauai (Hawaiian Islands) 228
kelp forests 198–9
Kilauea (Hawaii) 70–1
Knoop scale 42
Kuiper Belt 23
kyanite 38

L

La Palma (Canary Islands) 142–3
labradorite **50–1**
Ladakh Range 114

Lagerstätten **296–7**
lakes 62, 116, **164–5**, 224, 225
  meltwater **175**
  salt 81
  temperature layers **165**
lamellar twinning 50
land, life on 263, **274–5**
Landsats 109
landslides 107, 128–9, 132, 161
larvae 182
Late Heavy Bombardment 32
latent heat **101**
lateral moraines 170
lava
  basaltic 65, 111, 120
  continent formation 30
  ocean tectonics 206
  rock cycle 62, 63
  types of **72–3**
  viscosity **73**, 141
  volcanic eruptions 141
  volcanic islands 209
  volcano shapes 139
lava bombs 141
lava dome volcanoes 139
layering, forest **266**
layering, rocks 92-3
leaves 24, 25, 263, 306
  conifer 309
Leonid meteors 210
lichens 265
life
  biosphere **246–7**
  Cambrian explosion **256–7**
  cold conditions **304–5**
  dinosaurs **290–1**, **294–5**
  distribution of **302–3**
  DNA **310–1**
  drought conditions **280–1**
  early 32, **250–1**, 256
  on Earth 22, 246–7
  fish **260–1**
  flying animals **288–9**
  forests **266–7**, **308–9**, 314–5
  fossils **268–9**, **282–3**, 286–7, **296–7**
  grasslands **300–1**, **310–1**
  human impact **316–7**
  insect pollinators **292–3**
  on land **262–3**, **274–5**
  mammals **298–9**
  mass extinctions **276–7**, **294–5**
  megafauna **312–3**
  multi-celled organisms **254–5**
  in oceans 246
  origin of 16
  oxygen-fuelled giants **272–3**
  permafrost **306–7**
  reefs **258–9**
  species **248–9**
  and submarine volcanism 206
  succession **264–5**
  transforms Earth **252–3**
  trilobites **278–9**

life *continued*
 warm-blooded animals **284–5**
 water cycle 225
 see also animals; habitats
lift 288
light penetration, oceans 202
lightning 233, **236–7**, 238
lignin 270
lignite 271
limestone **82–3**, 161
 caves 168
 fossiliferous **82**
 inorganic 81
 karst 155, **166–7**, 168
 marine 124, 246
 organic 81, 251
lithification 79
lithium 12
lithosphere 107, 111, 112, 115,
 142, 247
lizards 297
loess **74**, 151
low Earth orbits 217
lungs, dinosaurs 290
lustre **40–1**
Lyell, Charles 196
Lystrosaurus 276

M

Maclure, William 85
macrocrystalline quartz 48
magma
 altered minerals 52
 contact metamorphism 90, 91
 continent formation 30
 gas bubbles 57
 igneous rocks 65, **70–1**, 135
 intrusions 54, 90, 135
 rock cycle 62, 63
 silica content 72
 tectonic plates 111, 116
 volcanic dykes 135, 137
 volcanoes 115, 139, 141, 142,
 206
magma chambers 62, 65, 67, 70,
 135, 139
magnesium 47
magnetic field, Earth's 20, 21,
 22, **213**
magnetic poles 213
magnetite 253
magnetosphere 213
malachite 39, 52–3, 54
mammals **298–9**
 ancestors 276
 body heat 2,84
 early 282–3, 285, 286, 298
 fossils 296–7
 grazing 300
 Great American Interchange **303**
 megafauna **312–3**
mammatus clouds 222, 223
mammoths, woolly 304–5, 311
manganese 47
mangrove swamps 195

mantle 30, 107
 igneous rocks 67, 107
 mountain roots 120
 subduction 115
 tectonic plates 111
 upper 65, 70, 107, 111, 112, 126
mantle plumes 139
mapping
 rocks **84–5**
 seafloor **200–1**
marble 82, 250–1
marcasite 36
Mariana Islands 209
marine organisms 81, 82, 179
marine reptiles 283
Mars 22–3
marsupials 298, 299, 312
Masai-Mara (Kenya) 300–1
mass extinctions 32, **276–7**,
 278, 294–5, 300
mass spectrometry 217
matrix (crystal growth) 54, 79
matter, states of 98
meanders **158**, 162
mechanical weathering 154
medial moraines 170
Mediterranean 179
megafauna **312–3**
Megaloceros giganteus 312–3
Megalonyx 303
Meganeura 272–3
Megatherium 302
melanosomes 311
meltwater 157, 158, 165,
 174–5, 224, 306
Mercury 23, 47
meroplankton 182
Merychippus 300
Mesohippus 300
mesolite 38
mesosphere 210, 222, 238
Mesozoic era 119
Messel Pit (Germany) **296–7**
metabolism 284, 290, 304,
 305
metals 44, **46–7**
metamorphic aureoles **91**,
 135
metamorphic rocks **86–91**
 burial **86–7**
 contact **90–1**
 dynamic **86–7**
 regional **88–9**
 rock cycle 62–3, 67
meteor showers 210
meteorites **16–7**, 22, 28, 50
meteoroids 20, 210
methane 184, 210, 241, 276
mica 47, 48
microbes 206, 249, 251, 252,
 253, 265
 reef building 258, 259
microcrystalline quartz 48
micrometeorites 16
microwaves 109, 217
Mid-Atlantic Ridge 116–7,
 201

mid-ocean ridges 65, 111, 116,
 117, 142, 179, 201, 205, 206
Midnight Sun 25
migrations 300
Milankovitch, Milutin 243
Milankovitch cycles **243**
Milky Way **13**
millipedes 272
minerals **36–61**
 altered **52–3**
 associations **54–5**
 crystal habits **38–9**
 crystal structure **36–7**
 feldspars **50–1**
 fossilization 268
 gemstones **58–9**
 hardness **42–3**
 hot springs and geysers 146
 inorganic 246, 247
 and life 246
 lustre **40–1**
 native elements **44–5**
 organic **60–1**
 sediment 205
 submarine volcanism 206
mist 222
Moho 67
Mohs scale 42
molecular clouds **13**
moles 299
molluscs 274, 294
moment-magnitude scale 128
monsoons 125, 195, 218
Monterey Bay (California, US)
 184
Montreal Protocol 217
Monument Valley (Utah, US) 137
Moon **20–1**, 24, 32
 meteorites 16
 rocks 50
 seismic activity 131
 tides 186
moons (planets other than Earth)
 14
moose 309
moraines 157, 170, 175
Morocconites 278
Mosellophyton 263
mosses 306
moulins 175
Mount Rainier National Park
 (Washington State, US) 152–3
mountains
 fold **118–9**, 120
 mountain building **120–5**
 oceanic 201
 plate boundaries 112, 115
 roots **120**
 uplift 196
mouths, river 158
mud 33, 162
mudpots 146
mudskippers 274
mudslides 132
mudstone 74, 286
multi-celled organisms **254–5**
mylonite 87

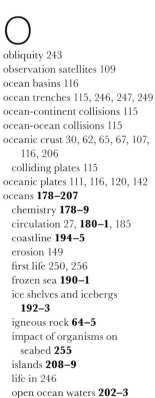

N

nacreous clouds 222
native elements **44–5**
native metals 47
Navajo sandstone 77
Nazca Plate 111
Neanderthals 310–1
neap tides 186
nectar 292
nematode worms 274
nematophytes 263
Neptune 23
nesosilicates 50
niches 298, 299
nickel 17, 47, 107
nimbostratus clouds 222
Nimbus-7 217
Nishinoshima (Japan) 208
nitrogen 28, 179, 210, 213, 241
noctilucent clouds 222
nocturnal animals 285
non-metals 44
normal faults 126
North American Plate 117, 172
North Pole 190, 213
northern lights (aurora borealis)
 212–3
Nostoc 253
notochord 260
nuclear fusion 15
nunataks 243
nutrients
 oceans 198, 202
 succession 265

O

obliquity 243
observation satellites 109
ocean basins 116
ocean trenches 115, 246, 247, 249
ocean-continent collisions 115
ocean-ocean collisions 115
oceanic crust 30, 62, 65, 67, 107,
 116, 206
 colliding plates 115
oceanic plates 111, 116, 120, 142
oceans **178–207**
 chemistry **178–9**
 circulation 27, **180–1**, 185
 coastline **194–5**
 erosion 149
 first life 250, 256
 frozen sea **190–1**
 ice shelves and icebergs
 192–3
 igneous rock **64–5**
 impact of organisms on
 seabed **255**
 islands **208–9**
 life in 246
 open ocean waters **202–3**
 origin of **28–9**, 32, 176
 pollution 317

oceans *continued*
  sea-ice **190–1**
  sea-level changes **196–7**
  seafloor **204–5**
  seafloor mapping **200–1**
  seafloor relief **184–5**
  shallow seas **198–9**
  tectonics **206–7**
  tides **186–7**
  tsunamis **128**
  upwelling and plankton
    blooms **184–5**
  volcanoes 139
  warming 314
  waves **188–9**
  zones 256
octopuses, dumbo 202
Okavango Delta (Botswana)
    162
Olenellus 278
oligoclase 51
olivine 17
omphacite 106, 107
Onnia 279
onyx 49
Oort Cloud 23
ophiolites 65, 208
orbits
  Earth's 10, 25, 243
  Moon's **21**
  Solar System **23**
orchids 266
Ordovician period 278, 279
organic matter 247, 250
organic minerals **60–1**
organic sedimentary rocks 81
Orion Spur 13
orthoclase 36, 51
osmium 44
overtopping 175
overturned folds 118, 119
overturning circulation **180**
oviraptorosaurs 281
oxygen
  atmospheric 210, 213, **272**
  and giant lifeforms **272–3**
  in ice cores 241
  and life 249
  photosynthetic 253
  sea water 179
ozone 109, 210, 217
ozone hole 217

P

Pacific Plate 142, 172, 208
pack ice 190
pahoehoe lava 72
palaeoanthropology 311
palaeontologists 283, 286, 311
Paleozoic era 278
Palmieri, Luigi 130, 131
pancake ice 190
Pangaea 124, 280
Paradoxides 278
parasitic folds 118

pearls 59, **60**, 61
peat 271, 314
pebbles 157
Pele's tears 70
pendulums 26–7
penetration twins 36
penguins, chinstrap 192–3
peridot 58
peridotite 65, 107
periodic table 47
permafrost 162, **306–7**, 308
Permian period 276, 278
permineralization 268
petals 292
Phanerozoic eon 32
Philippine Plate 208
phonolite porphyry 135
phosphorus 179
photosynthesis
  carbon dioxide 247, 271
  forests 266
  microbes 182, 251, 252, 253, 255
  multi-layered community 263
  oxygen generation 252, 253, 272
  plankton 204
  shallow seas 198
  and ultraviolet radiation 210
  water cycle 224
phyllosilicates 50
physical weathering 154
phytoplankton 182, 202
pillars, caves 168
pillow basalt 64–5
pillow lava 206
Pinatubo, Mount (Philippines) 139
pinnacles 167
pioneer species 265
pitcher plants 318
placental mammals 298, 299
placoderms 260
plagioclase feldspars 50
planetary differentiation 22
planetary winds 214
planetesimals 14, 17
planets, formation of **14**
plankton 204, 205, 256
  blooms **182–3**
plants
  biosphere 247
  deserts 281
  diversity of species 249
  fossilization 268, 269
  insect pollinators **292–3**
  on land **262–3**
  multi-layered community **263**
  oxygen production 253
  permafrost zones 306
  rainforest 318–9
  seasonal adaptation 24–5
  soil 97, 151
  transpiration 224
  water 98
  water cycle 224, 225
  see also forests
plasma 238
plastic pollution 317
platinum 44, 47

Pleistocene epoch 304, 305, 313, 317
Plinian eruptions 141
plumes, mantle 30, 142, 209
plunge pools 161
Pluto 23
plutons 135
podzols 97
polar ice sheets 306
pollen 24, 223
pollination, insect 248–9, **292–3**
pollution 317
  atmospheric 217
polymers 44
polymorphs 36
polyps, coral 258
porous rocks 70
porphyritic rocks 70
potassium-rich feldspars 50
potholes 149
pouches, marsupial 298
powder snow avalanches 102
precession 243
precipitation 167, 247
  rainfall **226–7**
  snow **230–1**
  thunderstorms **232–3**
  water cycle 224
  winter **226**
predators, absence of 303
prehnite 54–5
pressure
  air 27, 218, 219
  geysers **146**
  glaciers 170
  metamorphic rock 63, 87, 89, 107
  wind 214
prevailing winds 214, 215, 219
primary succession 265
primates 297
prismatic crystals 38, 39
Proetida 278
proglacial lakes 175
Proterozoic eon 32
protons 213
protoplanets 14, 22
Prototaxites 263
protozoans 205
pseudopodia 205
pterosaurs 288
Pulau Ndaa (Indonesia) 208–9
Putorana Mountains (Russia) 276–7
pyrite **36–7**, 54
pyroclastic flows 141
pyroclastic rocks 70
pyrophyllite 38
pyroxene 89, 106

Q

quaggas 311
quarrying, glacial 153, 170
quartz 36, 38, 40, 47, 53, 54, 77, 79
  geodes **56–7**
  varieties of **48–9**
quartz sandstone 77
Quaternary Ice Age 243

R

radar satellites 109
radiation belts 109
radiolarians 205
radiometric dating **28**
Rafflesia 318
Rainbow Mountains (China)
    92–3
raindrops **226**
rainfall **226–7**
  deserts 280, 281
  freezing 226
  high 228
  lakes 165
  salt cycle 179
  thunderstorms 233, 237
  tropical cyclones 220
  water cycle 224
rainforests 228, **266–7**, 318
rapids 158
ravines 92
Red Deer River Valley (Canada)
    286–7
Red Sea 179
reefs *see* coral reefs
reflection, crystals 41
refraction, crystals 41
regional metamorphism **88–9**,
    90
reproduction 249
  plants 263, 293
reptiles 280, 284, 296, 297
resin 61, 269, 292, 309
reverse faults 126
rhinoceroses 304
rhodochrosite 38
Rhynia major 262–3
rhyolite 73
Richter scale 128
rift valleys 116, 165
rivers **158–9**
  deltas and estuaries **162–3**, 195
  erosion and transport 62, 63, **149**
  lakes 165
  salt cycle 179
  sediment deposition **157**, 205
  underwater 198
  water cycle 224, 225
  waterfalls **160–1**
Roaring Forties 214
roches moutonnées **153**
rock crystal 48
rock cycle **62–3**, 67
rock flour 193
rocks
  age of Earth 32–3
  continents 30
  Moon 21
  planets 14
  primary succession 265
  see also igneous rocks; metamorphic
    rocks; sedimentary rocks
rodents 297
rogue waves 188
Rossby waves 216, 218

rotation
    Earth's 21, 25, **26–7**, 180, 182, 186, 218, 220, 243
    Moon's 21
rounding 153
ruby 58
rugose corals 258, 259

S

Sacco, Federico 123
salinity
    estuaries 162
    oceans 179, 180, 185, 202
salt crusts **81**
salt cycle **179**
salt pans 81, 179
saltation 149
San Andreas Fault 112–13
sand 33, 53, 150, 151, 157, 162
sand bars 157
sand dunes 151, 198
sandstone **76–7**, 126, 137, 161, 286
sandstorms 151
sap 309
sapphire 58
satellites **108–9**, 201, 217
Saturn 23
sauropods 291
Saussure, Horace Bénédict de 123
savannas 300
Sawyer Glacier (Alaska, US) 175
schools (fish) 202
scratches
    glacial 153, 170
    minerals 42
sea levels
    changes **196–7**
    lower 168
    rising 175, 314
sea stacks 149
sea water 162
sea-ice **190–1**
seafloor 198, **204–5**
    mapping **200–1**
    relief **184–5**
    spreading 116
seamounts 209
seas see oceans
seasons **24–5**
secondary minerals 52
secondary succession 265
sedges 306
sediment
    coal formation 271
    coastline 195
    continent formation 30
    deltas 162
    deposition 62–3, **156–7**, 162, 170
    diagenesis 53
    fossilization 268–9
    glaciers 175
    oceans 184, 198, 201, 206, 246
    rivers 149, 158
    seafloor 205, 254, 255
    stromatolites 251

sedimentary rocks **74–83**
    chemical deposition 63, **80–1**
    coal 271
    conglomerates **78–9**
    cross-bedding **77**
    fine-grained **74–5**
    formation of 157
    layers of 91, 93
    limestone **82–3**
    reef formation 258
    rock cycle 62–3, 67
    sandstone **76–7**
    wind erosion 151
seed plants 247, 292, 293
seiche waves 175
seismic waves 107, 128, 132
seismograms 131
seismographs 130, 132
seismometers 131
Selenopeltis 279
semi-metals 44, 46
serpentinite 86–7
serpentinization 86
sexual reproduction, plants 292, 293
Seymouria 274–5
shale 74, 90, 91
Shark Bay (Australia) 251
shear zones **87**
sheet lightning 237
shell debris 77, 205, 246
shells 60–1, 82, 256
shield volcanoes 139, 228
Shiprock (New Mexico, US) **136–7**
shockwaves 126
Siberia 276–7
Sierra Nevada Mountains (US) 68, 135
Silfra Fissure 117
silica 79, 205, 256
silicates 47, 48, 50, 252
    groups **50**
sills 135
silt 33, 74, 157, 162
siltstone 74
Silurian period 278
silver 44, 47
single-celled organisms 250, 254, 295
sinkholes 168
Sinosauropteryx 311
sinters 145
skarn 90
skeletal remains 77, 82, 179
Skeleton Coast (Namibia) 194–5
skeletons
    dinosaur 290–1, 294–5
    fish 260
    fossilization 268, 283
    mammals 285, 298–9
slab avalanches **102**
slate 88–9, 90
sleet 226
slip-off slopes 158
Slope Point (New Zealand) 214–5
slot canyons 77
sloths 303
sluffs 102
slugs 274

Smilodon 313
Smith, William 85
smoky quartz 48–9
snails 274
snakes 297
snow 226, **230–1**
    avalanches **102–3**
    glaciers 170, 193
    meltwater 175
    snowflakes 101, 230, **231**
    thunderstorms 233, 237
    water cycle 224, 225
snow line 247
Snowdonia National Park (Wales) 154–5
soda straws 168
soil 34, **96–7**, 247
    above permafrost 306
    liquification 132
    profiles **97**
    succession 265
solar energy 25, 180, 182, 218, 225, 246
solar radiation 243
Solar System 10, 13, **14–5**, 16
    Earth's place in **22–3**
solar wind 20, 22, 213
solstices 24
sonar 201
sonic shock waves 237
Sorge, Ernst 240
sorosilicates 50
South Pole 190, 213
southern lights (aurora australis) 213
space dust 205
space, edge of 210
species
    evolution through geographical separation **249**
    variety of **248–9**
speleothems 168
sphalerite 54
spiders 274
spin, Earth's 234
spinal cord 260
spinel 36
spiral galaxies 13
sponges 258, 259
spores 223, 247, 263
spreading ridges 117
Spriggina 254
spring tides 186
springs 158, 168, 224
sprites, red 238–9
stacks, sea 195
stalactites 167, **168**
stalagmites 167, **168**
stamens 292
stars 12–3
static electricity 237
steam 98, 145, 146
steel 47
stems, plant 263
steppes 300, 304, 308
stony-iron meteorites 17
storm surges 220

storms
    Coriolis effect 27
    ocean 188, 202
    thunderstorms **232–3**
    tropical cyclones **220–1**
strata 33, **92–3**
stratocumulus clouds 222
stratosphere 210, 238
stratovolcanoes 139, 141
stratus clouds 223
striations 37, 170
strike-slip faults 126
Strokkur Geyser (Iceland) 146–7
stromatolites 60, **250–1**, 252, 258, 259
Strombolian eruptions 141
structure, Earth's **106–7**
Studlagil Canyon (Iceland) 63
subduction/subduction zone 30, 66, 111, 115, 208, 209
submarine fans 198, 201
subsoil 97
succession **264–5**
Suess, Eduard 122, 123
sulphur 44, 46
Sumatran trench 201
Sun **14–5**
    Earth's orbit 10, 25, 243
    Solar System orbits **23**
    ultraviolet radiation 210
sunlight 24, 25, 246, 247
supercells 233
Superior, Lake (Canada/US) 165
supernovae 12
surf 188
surface currents 180
surface tension **98**
surface waves 128, 132
suture zone 115
swell 188
symmetry, crystals 36, 38
synchrotron particle accelerators 283
synclines 118, 119

T

taiga forests **308–9**
Taklamakan Desert (China) 156–7
talc 42
tapirs 297
Tasmania (Australia) 148–9
technology 317
tectonic activity 57, 93
tectonic lakes 165
tectonic plates **110–1**
    boundaries 65, 87, 111, **112–3**, 115, 119, 120, 128, 139, 146
    colliding 89, 94, 112, **114–5**, 120, 172, 208
    continent formation 30
    convergence 62, 107, 112, 120
    divergence 112, **116–7**, 165
    folds 119
    hotspot volcanoes 142
    island formation 208
    mountain building **120–1**
    oceans 201, **206–7**

tectonic plates *continued*
  rock cycle 16, 62–3
  sea levels 196
  subduction 30, 66, 111, 115, 208, 209
  transform boundaries **112**
  and water 224, 225
tectosilicates 50
teeth
  browsers and grazers 300
  dinosaurs 290, 291, 295
  fossils 282, 283
  mammals 299
Tempel-Tuttle comet 210
temperature
  animals **284**
  boreal forests 308
  geothermal features **144–5**, 146
  lakes **165**
  metamorphic rock 63, 87, 89
  natural climate change 243
  ocean 314
  snowflake morphology 231
  water 98, 101
terminal moraines 170
terminus, glacial 175
tests 205
tetrahedrite 46
tetrapods 274
Tharp, Marie 201
thermocline 165, **202**
thermohaline conveyor belt 184, **185**
thermosphere 210
theropods 290
Thrinaxodon 282–3
thrust (flight) 288
thrust faults 126
thunder 233
thunderstorms 220, **232–3**, 238
Thursophyton 263
Thylacoleo carniflex 298–9
Tibetan Plateau 125
tides 21, 131, 162, **186–7**, 198
tidewater glaciers 172
Tien Shan Mountains (China) 126–7
tiger eye 253
tigers 313
tilt, Earth's 21, **24**, 25, 243, 304
tilted layers 92–3
Titanis walleri 302–3
titanium 47
Titanopodus 291
Tnorala (Australia) **18–9**
Tonga 209
topaz 58
topsoil 96, 97
torbernite 39
Tornado Alley (US) **234–5**
tornadoes 233, 234
tourmaline 58
trace fossils 268
traction 149
trade winds 215, 228
transform boundaries **112**
Transient Luminous Events (TLEs)
  238–9
transpiration 224
traps 276

trees **266–7**
  coal formation 270, 272
  conifers 308, **309**
  seasonal adaptation 24–5
  transpiration 224
  see also forests; rainforests
tremolite 38
trenches, ocean 115, 246, 247, 249
Triarthrus 279
tributaries 158
Triceratops 294–5
trilobites 256–7, **278–9**
Tristichopterus 260
tropical cyclones **220–1**
troposphere 210, 214, 220, 222, 233
trunks, tree 266, 270
tsunamis **128**, 184, 201
tundra 306, 308
Tungurahua Volcano (Ecuador) 141
Turakirae Head (New Zealand) 196–7
turbidity currents 184, 198
Turnagain Arm (Alaska, US) 186–7
turquoise 59
turtles 297
tusks 304–5
typhoons 220
tyrannosaurs 286
Tyrannosaurus rex 290–1, 311

UV

U-shaped valleys 152–3
ulexite 40
ultraviolet radiation 109, 210
uniformitarian theory of geology 196
Universe, birth of 12
updrafts 226, 233
uplift 62, 196
upwellings **182–3**
  coastal 198
  open ocean 202
uranium 28
Uranus 23
vascular plants 265
vents, volcanic 139
Venus 23
vertebrates
  fish 260–1
  on land 274–5
  mammals 298–9
  Messel Pit fossils 296–7
vertical diffusion **182**
vesicles 21
Vesuvius, Mount (Italy) 131
Victoria, Lake 165
vines 266
Viviani, Vincenzo 26
voids, in rocks 57
volcanic islands **142**, 208, **209**,
  303
volcanic lightning 236–7
volcanoes **138–9**
  and atmosphere 210
  carbon dioxide emissions 271
  colliding plates 115
  early Earth 22

volcanoes *continued*
  eruptions **140–1**
  extinct 137, 228
  geothermal features **144–5**, 146
  hotspots **142–3**, 209
  ice core records 241
  igneous rocks 70
  island chains **142**
  lightning 236–7
  mountain building 120
  Permian eruption **276**
  plate boundaries 112
  rock cycle 62, 63
  salt cycle 179
  sediment 205
  subduction 66
  submarine 201, 206, 209
  succession 265
  types of 139
Vulcanian eruptions 141

W

Wai'ale'ale, Mount (Kauai) **228–9**
Walliserops 278–9
walruses 193
wapitis 313
warm-blooded animals **284–5**,
  296–7, 312
wasps 297
water
  deltas and estuaries 162
  drought **280–1**
  erosion by 95, **148–9**
  extra-terrestrial 28
  ice **100–1**
  karst landscapes 167
  life on Earth 246
  Mars 23
  meltwater **174–5**
  ocean chemistry **178–9**
  properties of **98–9**
  rainfall **226–7**
  sediment transport/deposition 157
  soil 97
  surface 28
  see also glaciers; lakes; oceans; rivers
water cycle 145, 218, **224–5**
water table 167, 168, 224
water vapour 28, 98, 101, 210, 223,
  224, 226, 230, 247
waterfalls 69, **160–1**, 228
wave trains 188
waves
  erosion 63, 149, 195
  formation **188**
  lakes 165
  oceans **188–9**, 198
  seiche 175
  tides 186
weather satellites 109, 217
weather systems 210, **218–9**
weathering 30, 66, 79, 92, 104, **154–5**
  rock cycle 62, 63
  see also erosion
wet snow avalanches 102

White Mountains (New Hampshire,
  US) 135
wildebeest, blue 300–1
wilderness 317
wildfires 265, 272
wind 180, **214–5**, 218, 219
  speed 216, 217
  tropical cyclones **220–1**
  wave formation 188
wind abrasion 150
wind cells 214, 215
wind erosion **150–1**
wind pollination 293
wings 288, 289

XY

X-ray technologies 283
xylem vessels 263, 266
yellow dwarf stars 15
Yellowstone National Park (US) 142
Yosemite Falls (California, US) 69
Yosemite Valley (California, US)
  68–9, 135
Yucatan (Mexico) 168–9
Yukon Delta (Canada) 162–3
Yunnan Province (China) 155

Z

Zenapsis 260
zinc 47
zircon 28
Zlichovaspis 278
zooplankton 182, 202, 205

acknowledgments

DK would like to thank:
Lee Skoulding at My Lost Gems, Southwold, UK (mylostgems.com) for help with photoshoots; Ina Stradins for her contribution to design; Gary Ombler for photography; Steve Crozier for image retouching; Peter Bull for the artwork of the water cycle on p.224; ETH Zurich for providng new images of mountain models; Maya Myers for fact-checking; Aarushi Dhawan, Kanika Kalra, Arshti Narang, and Pooja Pipil for design assistance; Vijay Kandwal, Nityanand Kumar, and Mohd Rizwan for high-res assistance; Mrinmoy Mazumdar for DTP assistance; Ahmad Bilal Khan and Vagisha Pushp for picture research assistance; Rakesh Kumar for DTP design on the jacket; Tom Booth for the glossary; Richard Gilbert for proof-reading; and Helen Peters for compiling the index.

The publisher would like to thank the following for their kind permission to reproduce their photographs:

(Key: a-above; b-below/bottom; c-centre; f-far; l-left; r-right; t-top)

1 Dreamstime.com: Daniel127001. **2 Science Photo Library:** Eye Of Science. **4-5 Getty Images:** Arctic-Images. **8-9 Popp-Hackner Photography OG:** Verena Popp-Hackner & Georg Popp. **10-11 123RF.com:** Roberto Scandola. **12-13 NASA:** ESA / N. Smith (University of California, Berkeley / The Hubble Heritage Team. **13 NASA:** CXC / SAO / JPL-Caltech / STScI (crb). **14 ESO:** ALMA (ESO / NAOJ / NRAO) (tl). **14-15 ©Alan Friedman / avertedimagination.com**. **16**

Getty Images: Walter Geiersperger / Corbis Documentary (tl). **Science Photo Library:** Dennis Kunkel Microscopy (cl). **16-17 Getty Images:** Walter Geiersperger / Corbis Documentary (b). **18 Shutterstock.com:** Auscape / UIG (crb). **18-19 Alamy Stock Photo:** Jean-Paul Ferrero / AUSCAPE. **20 NASA**. **21 Shutterstock.com:** Michael Wyke / AP (ca). **22 Alamy Stock Photo:** Michael Runkel / robertharding (tr). **22-23 ESA:** DLR / FU Berlin. **25 Alamy Stock Photo:** Composite Image / Design Pics Inc (br). **26-27 Robert Harding Picture Library**. **27 Alamy Stock Photo:** B.A.E. Inc. (tr). **28-29 ESA:** Rosetta / NAVCAM. **28 J. W. Valley, University of Wisconsin-Madison:** (cl). **30 Courtesy of Smithsonian. ©2020 Smithsonian:** Chip Clark, NMNH (tl). **31 John Cornforth. 32-33 Jeffrey Sipress. 34-35 Getty Images:** mikroman6 / Moment. **37 Alamy Stock Photo:** Phil Degginger. **38 Alamy Stock Photo:** J M Barres / agefotostock (cra); Natural History Museum, London (cr). **Depositphotos Inc:** Minakryn (fbl). **Dreamstime.com:** Annausova75 (cl). **Science Photo Library:** Phil Degginger (fcr); Charles D. Winters (cla); Natural History Museum, London (bl); Dirk Wiersma (br); Millard H. Sharp (fbr). **Shutterstock.com:** Albert Russ (fcra). **39 Dorling Kindersley:** Holts Gems (cl). **Dreamstime.com:** Bjrn Wylezich (bl). **Shutterstock.com:** Albert Russ (r). **40-41 Shutterstock.com:** Bjoern Wylezich. **40 Alamy Stock Photo:** J M Barres / agefotostock (cb); J M Barres /

agefotostock (crb). **Dreamstime.com:** Miriam Doerr (bl); Infinityphotostudio (clb). **Shutterstock.com:** Sebastian Janicki (br). **43 Petra Diamonds. 44 Alamy Stock Photo:** Andreas Koschate / Westend61 GmbH (ca). **Dorling Kindersley:** RGB Research Limited (clb). **Dreamstime.com:** Bjrn Wylezich (cla); Bjrn Wylezich (cb). **44-45 Alamy Stock Photo:** Bjrn Wylezich. **46-47 Crystal Classics LTD / crystalclassics.co.uk**. **47 Dreamstime.com:** Bjrn Wylezich (c). **48 Dreamstime.com:** Daniel127001 (fcla); Bjrn Wylezich (cla); Epitavi (fcl). **Science Photo Library:** Millard H. Sharp (fcr). **Shutterstock.com:** Sebastian Janicki (cl); Sebastian Janicki (cr). **49 Dorling Kindersley:** Natural History Museum, London (cra). **Getty Images:** Walter Geiersperger / Corbis Documentary (cla). **Science Photo Library:** Joyce Photographics / Science Source (ca). **Shutterstock.com:** Sebastian Janicki (b). **50-51 Shutterstock.com:** Sebastian Janicki. **51 Alamy Stock Photo:** Susan E. Degginger (cra); Susan E. Degginger (fcra). **Science Photo Library:** Phil Degginger (fcla). **Shutterstock.com:** Henri Koskinen (cla). **52-53 Dreamstime.com:** Bjrn Wylezich. **54 Dreamstime.com:** Bjrn Wylezich (cl). **56 Alamy Stock Photo:** Roland Bouvier (bl). **56-57 Dreamstime.com:** KPixMining. **58 Alamy Stock Photo:** Enlightened Media (clb); Valery Voennyy (cla). **Science Photo Library:** Paul Biddle (crb); Mark A. Schneider (cb); Charles D. Winters (bl). **Shutterstock.com:** Cagla Acikgoz (bc); DmitrySt (ca);

luca85 (cra); Minakryn Ruslan (br). **59 Alamy Stock Photo:** Ian Dagnall (cla); Natural History Museum, London (br). **Dreamstime.com:** Tatiana Neelova (cra). **Shutterstock.com:** Jirik V (ca). **60-61 Dreamstime.com:** Cristian M. Vela. **61 Depositphotos Inc:** Dr.PAS (cla). **Dreamstime.com:** Avictorero (cra). **Shutterstock.com:** valzan (ca). **64-65 Getty Images:** Dale Johnson / 500Px Plus. **65 Dreamstime.com:** Michal Baranski (bc). **66-67 Dreamstime.com:** Lukas Bischoff. **67 Dreamstime.com:** Zelenka68 (tl). **68-69 Unsplash:** Robby McCullough. **69 Unsplash:** Yang Song (cb). **70 Shutterstock.com:** MarcelClemens (tl). **70-71 Dreamstime.com:** Zelenka68. **72-73 Bryan Lowry**. **72 Alamy Stock Photo:** tom pfeiffer (clb). **74-75 Richard Bernabe Photography**. **76-77 Getty Images:** Piriya Photography. **77 Shutterstock.com:** Yanping wang (br). **78-79 Dreamstime.com:** . **79 Alamy Stock Photo:** J M Barres / agefotostock (bc). **80-81 The Light Collective:** Ignacio Palacios. **81 Alamy Stock Photo:** Natural History Museum, London (tc). **82 Dreamstime.com:** Losmandarinas (bl). **82-83 Alamy Stock Photo:** Natural History Museum, London. **84 British Geological Survey:** Permit Number CP22 / 005 British Geological Survey © UKRI. **85 Alamy Stock Photo:** Natural History Museum, London (cl). **Trustees of the Natural History Museum, London:** (t). **87 Alamy Stock Photo:** Siim Sepp (tr). **88-89 Getty Images:** Pete Rowbottom / Moment. **89**

Alamy Stock Photo: J M Barres / agefotostock (tc). **90 Alamy Stock Photo:** Siim Sepp (cra). **90-91 Getty Images:** Dean Fikar / Moment. **92-93 Getty Images:** MelindaChan / Moment. **93 Science Photo Library:** Sinclair Stammers (br). **94-95 Getty Images / iStock:** Meinzahn. **95 Getty Images / iStock:** Pears2295 (crb). **96-97 Alamy Stock Photo:** michal812. **97 Alamy Stock Photo:** kristianbell / RooM the Agency (br). **98 Getty Images:** Sherry H. Bowen Photography / Moment (tl). **98-99 Alamy Stock Photo:** Dennis Hardley. **100-101 Getty Images:** Sergey Pesterev / Moment. **101 Alamy Stock Photo:** Guy Edwardes Photography (br). **102-103 Paddy Scott. 104-105 Alamy Stock Photo:** Quagga Media. **106-107 Alamy Stock Photo:** J M Barres / agefotostock. **107 Alamy Stock Photo:** Kitchin and Hurst / All Canada Photos (cl). **108 Alamy Stock Photo:** NASA / agefotostock. **109 NASA. 110-111 Getty Images:** Posnov / Moment. **111 Alamy Stock Photo:** van der Meer Marica / Arterra Picture Library (tr). **112 Alamy Stock Photo:** Susan E. Degginger (bl). **112-113 Getty Images:** DEA / Pubbli Aer Foto / De Agostini Editorial. **114 Alamy Stock Photo:** Hitendra Sinkar. **116 Shutterstock.com:** Joanna Rigby-Jones (tr). **116-117 Getty Images:** by wildestanimal / Moment. **118-119 Science Photo Library:** Bernhard Edmaier. **119 Alamy Stock Photo:** Matthijs Wetterauw (br). **120-121 Roberto Zanette. 122-123 Department of Earth Sciences, ETH Zrich. 123 Science Photo Library:** Eth-Bibliothek Zrich (tr). **124-125 EyeEm Mobile GmbH:** Salvatore Paesano. **125 Getty Images:** iGoal.Land.Of. Dreams / Moment (crb). **126-127 NASA:** Earth Observatory images by Robert Simmon and Jesse Allen, using Landsat data from the USGS Earth Explorer.. **126 Dreamstime.com:** Chris Curtis (bl). **128-129 Shutterstock.com:** Jiji Press / Epa-Efe. **131 Science Photo Library:** NASA (cl). **132-133 Anchorage Daily News:** Marc Lester. **132 ESA:** Copernicus data (2014) / ESA / PPO.labs / Norut / COMET-SEOM Insarap study (bl). **134-135 Getty Images / iStock:** RiverNorthPhotography / E+. **135 Dreamstime.com:** Sarit Richerson (cla). **136-137 Getty Images:** Wild Horizon / Universal Images Group Editorial. **137 Alamy Stock Photo:** Brad Mitchell (clb). **138-139 James Rushforth. 139 Getty Images:** Ignacio Palacios / Stone (cla). **140-141 Getty Images:** Daiva Baa / EyeEm. **141 Getty Images:** Sebastin Crespo Photography / Moment (cra). **142-143 NASA:** NASA Earth Observatory images by Lauren Dauphin, using Landsat data from the U.S. Geological Survey. **144-145 Getty Images:** Tobias Titz / fStop. **146 SuperStock:** Colin Monteath / age fotostock (tl). **146-147 Getty Images:** Arctic-Images / Stone. **148-149 Getty Images:** Monica Bertolazzi / Moment. **149 Alamy Stock Photo:** Art Publishers / Africa Media Online (tr). **150-151 Getty Images:** Agnes Vigmann / 500px. **151 Science Photo Library:** NASA (tr). **152-153 Alamy Stock Photo:** Dennis Frates. **153 Alamy Stock Photo:** blickwinkel / McPHOTO / TRU (tr). **154-155 Alamy Stock Photo:** Christopher Drabble. **154 Getty Images:** Walter Bibikow / DigitalVision (tr). **156-157 Alamy Stock Photo:** B.A.E. Inc.. **157 Science Photo Library:** Dr Morley Read (cr). **158 Getty Images:** Maxim Blinov / 500px Prime (tl).

158-159 Getty Images: Daniel Bosma / Moment. **160-161 Shutterstock.com:** travelwild. **161 Alamy Stock Photo:** Daniel Korzeniewski (tr). **162 Getty Images:** Martin Harvey / The Image Bank (cl). **162-163 NASA:** Jesse Allen and Robert Simmon / United States Geological Survey / Landsat 7 / ETM+. **164-165 ESA:** Copernicus Sentinel (2021). **165 Getty Images:** Posnov / Moment (tr). **166-167 Shutterstock.com:** bengharbia. **168 Getty Images:** Amrish Aroonda Manikoth / EyeEm (cl). **168-169 Getty Images:** Westend61. **170-171 Getty Images:** Bernhard Klar / EyeEm. **172-173 Getty Images:** Ignacio Palacios / Stone. **172 Dreamstime.com:** Bennymarty (crb). **174-175 Georg Kantioler. 175 Alamy Stock Photo:** Danita Delimont Creative (tr). **176-177 Dreamstime.com:** Channarong Pherngjanda. **178-179 Tom Hegen GmbH. 179 Getty Images / iStock:** mantaphoto (tr). **180-181 Science Photo Library:** Karsten Schneider. **182 Alamy Stock Photo:** Doug Perrine / Nature Picture Library (bc); Eric Grave / Science History Images (cb); Scenics & Science (fcrb). **naturepl.com:** Shane Gross (br); Solvin Zankl (fbr). **Science Photo Library:** Wim Van Egmond (crb). **183 Science Photo Library:** NASA. **184 Science Photo Library:** Ryan Et Al / Geomapapp (cra). **184-185 NASA:** Image courtesy Serge Andrefouet, University of South Florida.. **186 Alamy Stock Photo:** Paul Brady (cla). **186-187 Getty Images:** Streeter Lecka. **188-189 Josef Valenta Photography. 188 Flying Focus aerial photography:** (bl). **190-191 Getty Images:** Cavan Images. **190 Getty Images:** Fuse / Corbis (tl). **192-193 Getty Images:** Holger Leue / The Image Bank. **193 Getty Images:** Paul Souders / DigitalVision (tr). **194-195 Dreamstime.com:** Rabor74. **195 Science Photo Library:** Planetobserver (br). **196-197 Rob Suisted / Nature's Pic Images. 196 Getty Images:** Photo 12 / Universal Images Group Editorial (bc). **Julian Hodgson:** (tl). **198-199 Jon Anderson Wildlife Photography. 198 naturepl. com:** Franco Banfi (bl). **200 Library of Congress, Washington, D.C.:** Berann, Heinrich C. / Heezen, Bruce C. / Tharp, Marie. **201 Alamy Stock Photo:** Granger - Historical Picture Archvie (cl). **Library of Congress, Washington, D.C.:** Berann, Heinrich C. / Heezen, Bruce C. / Tharp, Marie (tr). **202 Alamy Stock Photo:** NOAA (tl). **202-203 Shutterstock.com:** Lovkush Meena. **204 Science Photo Library:** Steve Gschmeissner. **205 Science Photo Library:** Eye Of Science (cla); Eye Of Science (ca); Eye Of Science (cra); Eye Of Science (clb); Eye Of Science (cb); Eye Of Science (crb). **206-207 MARUM- Center for Marine Environmental Sciences, University of Bremen:** CC BY 4.0. **206 Science Photo Library:** OAR / National Undersea Research Program (tl). **208-209 Dreamstime.com:** Fabio Lamanna. **208 Getty Images:** DigitalGlobe / ScapeWare3d / Maxar (tr). **210 Science Photo Library:** Tony & Daphne Hallas (cra). **210-211 Shutterstock.com:** studio23. **212-213 Getty Images:** Elena Pueyo / Moment. **213 Shutterstock.com:** Vladi333 (br). **214-215 G Sharad Haksar:** Eye-light Pictures. **215 Alamy Stock Photo:** JordiStock (crb). **216 Alamy Stock Photo:** Science History Images. **217 NASA:** Albert J. Fleig, Jr. (cl); Robert Simmon (t). **218 Science**

The SCIENCE of ...

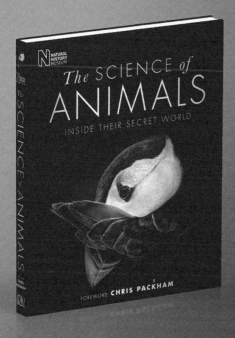

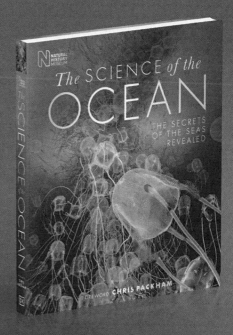

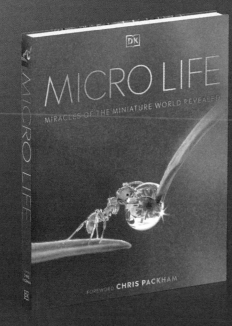

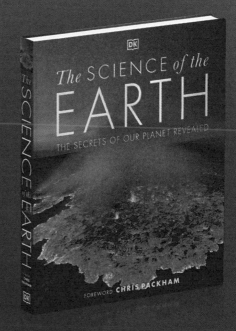

DK For the curious